# My Favorite Universe

## Professor Neil deGrasse Tyson

THE TEACHING COMPANY ®

## PUBLISHED BY:

## THE TEACHING COMPANY
4151 Lafayette Center Drive, Suite 100
Chantilly, Virginia 20151-1232
1-800-TEACH-12
Fax—703-378-3819
www.teach12.com

ISBN 1-56585-694-5

# Neil deGrasse Tyson, Ph.D.

The Frederick P. Rose Director, Hayden Planetarium,
American Museum of Natural History,
and Visiting Research Scientist and Lecturer, Princeton University

Neil deGrasse Tyson was born and raised in New York City, where he was educated in the public schools through his graduation from the Bronx High School of Science. Tyson went on to earn his B.A. in Physics from Harvard and his Ph.D. in Astrophysics from Columbia University.

Tyson's professional research interests include star formation, exploding stars, dwarf galaxies, and the structure of our Milky Way. Tyson obtains his data from telescopes in California, New Mexico, Arizona, and the Andes Mountains of Chile.

In addition to dozens of professional publications, Dr. Tyson has written, and continues to write, for the public. Since January 1995, he has written a monthly essay for *Natural History* magazine under the title "Universe." Tyson's recent books include a memoir, *The Sky Is Not the Limit: Adventures of an Urban Astrophysicist*; the companion book to the opening of the new Rose Center for Earth and Space, *One Universe: At Home in the Cosmos* (coauthored with Charles Liu and Robert Irion), which won the AIP science writing prize for 2001; and a playful question-and-answer book on the universe for all ages, titled *Just Visiting This Planet*. Also, premiering in the fall of 2004 will be a four-part PBS-NOVA special on *Cosmic Origins*, hosted and narrated by Tyson.

Tyson's contributions to the public appreciation of the cosmos have recently been recognized by the International Astronomical Union in its official naming of asteroid "13123 Tyson." On the lighter side, Tyson was voted "Sexiest Astrophysicist Alive" in the November 14, 2000, issue of *People Magazine*, the publication's annual "Sexiest Man Alive" issue.

Tyson is the first occupant of the Frederick P. Rose Directorship of the Hayden Planetarium, and he is a Visiting Research Scientist in Astrophysics at Princeton University, where he also teaches. Tyson lives in New York City with his wife and two children.

# Table of Contents

# My Favorite Universe

# My Favorite Universe

**Scope:**

This series of lectures discusses 12 topics based on 12 handpicked essays out of a hundred or so written for *Natural History* magazine since 1995. Although they do not follow a particular curriculum, they nonetheless represent the professor's favorite cosmic subjects. And, not surprisingly, they represent topics for which the general public harbors a sustained and insatiable interest.

The dozen lectures are thematically arranged in four groups of three. The first group might be entitled "On Being." Here, Professor Tyson introduces the fundamental properties of matter and energy and the forces that shape the cosmos. Describing these properties and forces as though they are protagonists on a cosmic stage, Tyson shows how the same laws of physics discovered here on Earth reveal themselves elsewhere in the universe, lending extraordinary confidence to the enterprise of science.

The next group of three lectures comes under the heading "Cosmic Catastrophes." Here, Professor Tyson highlights a battery of destructive cosmic phenomena and the role catastrophe has played in the history of life on Earth and in the history of Earth as a planet. The lectures include detailed descriptions of all the things that are bad for you, including black holes, the death of the Sun, and killer asteroids.

The next group of three lectures might be called "The Big Bang." For these lectures, Professor Tyson examines the frontier of our understanding of the universe and asks our most basic questions: How did our universe get here? How has it evolved in the past, and how will it evolve in the future?

Finally, the last group of three lectures addresses the most intriguing quest of them all: "The Search for Life in the Cosmos." Does life exist elsewhere? In what environments would we expect to find life? What would that life be like? Would we recognize alien life if we saw it?

The mission of the dozen lectures in *My Favorite Universe* is to pique your interest in some of the most fascinating and fundamental questions ever asked—questions that have been with us across time and across cultures. In the end, we will know that we have succeeded when "my favorite universe" becomes "your favorite universe."

# Lecture One
# On Being Round

**Scope:**

Let us begin by describing the property of "roundness." What forces tend to shape objects into roundness, and why is a sphere the most efficient shape that objects can take? From our discussion of spheres in nature on Earth, we move to spheres in the cosmos. Some planets are perfect spheres, but others are not, which in itself tell us something about their environments. As you will see, our description of roundness will take us across the cosmos.

## Outline

I. Why are so many things in the universe round?

　A. The forces that make things round operate on small and large scales.

　B. The term *round* refers to the energy of a body. Energy tends to descend to the lowest energy state it can; for example, think of a house of cards.

　C. Some things in the universe, such as crystals, are not round; this fact also tells us something about these objects.

II. Many natural objects, however, are round, such as soap bubbles, stars, planets, and galaxy halos. Even the observable universe is a perfect sphere, centered on us.

　A. This "roundness" is the result of forces that want to shape an object in such a way that the surface is minimized. Think, again, of soap bubbles. No matter the cavity through which you blow the soapy liquid, what comes out the other side is a sphere.

　　1. The sphere is the shape that encloses the largest volume with the least surface.

　　2. If the bubble were any other shape, it would have to stretch itself to cover the surface area. Any other shape would not be as strong as a sphere; it would be thinner in one place than another, and the bubble would pop.

**B.** This generalized feature is also revealed in a cube.

1. Some parts of the cube are more distant from the cube's center than others. Every corner is farther from the center than the middle of the cube's sides, which weakens the sides.

2. If the cube were an orb that had gravity—like a planet—the corners would be mountains, and the forces that had enabled gravity to make the planet in the first place would tend to make those mountains smaller. A rock on the mountain would roll down and fill up the center of the orb.

3. This process would continue until the cube much more closely resembled a sphere.

**C.** Other spheres in the universe include raindrops, which are not really tear-shaped but perfect spheres.

1. The force that holds a raindrop together is surface tension—the boundary between a liquid and the air.

2. In forming that boundary, the molecules of the liquid grab onto each other to establish the surface. The act of establishing the surface creates a tension that wraps the liquid. When it falls, the raindrop wraps itself into a perfect sphere, once again, making itself the most efficient shape that it possibly can.

**D.** Another perfect sphere is a ball bearing, but how is one made? It can't be produced with a lathe, because it is too small.

1. Ball bearings can be made by dropping liquefied metal down a tube. As the metal travels down the tube, it cools and hardens into a perfect sphere.

2. If you were in a weightless atmosphere, such as the space station, you could squeeze the liquefied metal from an eyedropper, and it would cool and harden and form a perfect sphere right in front of you. In fact, in a weightless atmosphere, you could produce the most perfect ball bearings ever made.

3. Mercury is the only metal that is liquid at room temperature, but its surface tension is so high that it forms a sphere under normal conditions on Earth. Think

of the toys that children used to play with in which a mercury bead traveled through a maze.

E. If a sphere maximizes volume and minimizes surface area—in other words, given that a sphere is the most efficient shape—why isn't everything a sphere? Why not packaging in the grocery store, for example? The answer is that spheres roll.

III. As we know, spheres also exist in the solar system.
   A. The Sun, which is a star, is a perfect sphere of gas. All the gas in the Sun tends to get as close to the center of gravity as possible to minimize how much total energy is expressed in that field of gravity.
   B. Saturn is one-tenth the size of the Sun and is a slightly flattened sphere. The fact that Saturn is not a perfect sphere tells us something about what's going on in Saturn's environment.
   C. The Earth, another sphere, is one-tenth the size of Saturn; the Moon is one-fourth the size of Earth. Gravity transforms all these objects of different sizes into spheres.
   D. Does gravity ever fail in its attempt to turn things into spheres? Yes.
      1. When an object is small and its field of gravity is weak, it will not become a sphere.
      2. Phobos, a moon of Mars that is one-tenth the size of our moon, is not spherical. Phobos does not have enough gravity to have wrapped itself into a sphere. Gaspra, an asteroid that is one-tenth the size of Phobos, is also not a sphere.
      3. The chemical bonds of the elements that make up these objects are stronger than the force of gravity, and gravity is helpless in its attempts to turn these objects into spheres. Our own bodies serve as another example.
   E. You might note that some of these objects that we have been referring to as perfect spheres do not seem "perfect" to us.
      1. Earth, for example, has craters, cliffs, valleys, and mountains.
      2. Keep in mind, however, that the deepest part of Earth's crust, the Marianas Trench, is 35,000 feet, or about six

miles, down. The highest point on Earth's crust, Mount Everest, is about 29,000 feet, or about five or six miles, up. The total distance, then, between the deepest and the highest points on Earth's surface is 12 miles.

3. This fluctuation is 1/600 of the diameter of the globe. If the Earth were shrunk down to the size of a cue ball, it would be absolutely smooth—a perfect sphere.

**IV.** Why are some things in the universe not spheres?

   **A.** Tidal forces pull some objects out of a spherical shape. The side of an object that is closer to the force of gravity will feel more gravity than the other side and will be pulled in the direction of the gravity.

      **1.** The Moon exerts tidal forces on Earth. The oceans respond to the fact that one side of the Earth is closer to the Moon than the other. The oceans on the closer side bulge out, resulting in high tide. The oceans on the other side also bulge but to a lesser degree. The oceans on the perpendicular sides experience low tides.

      **2.** The same tidal forces can be seen in binary stars, where two stars orbit each other. If one of these stars is a black hole and one is a blue or red supergiant, the tidal forces can become so great that some of the material from the giant will be funneled toward the black hole. As the black hole fills up, the side of the giant closer to the black hole feels an extra tug and its shape becomes distorted. Ultimately, it will resemble a Hershey Kiss.

      **3.** If one object comes too close to another, tidal forces can rip the first object apart. In the case of Saturn, an asteroid or comet came too close to the planet, and Saturn's gravity ripped it apart and scattered its material into a ring. Eventually, the particles of this asteroid or comet will fall out of orbit, and Saturn will lose its ring.

   **B.** Rotation also affects the shape of objects.

      **1.** In a rotating object, the movement of rotation will begin to collapse, and the object will be affected by what physicists call the *conservation of angular momentum*.

      **2.** This principle states that if an object is big and rotating slowly, as it gets smaller, it will compensate for getting smaller by speeding up. We see this principle in a skater who is spinning on an ice rink.

3. We see this same phenomenon in gas clouds. The rotation of the cloud preserves the plane, but the cloud itself collapses from top to bottom. The rotation has the effect of flattening the system.

4. This general flattening is also seen in galaxies. In the Milky Way, for example, some stars reveal the skeleton of the sphere that originally existed, but the galaxy has flattened out.

5. Earth, too, is slightly bigger at the equator than at the poles, because it is rotating at the rate of 25,000 miles around each day.

C. What happens if an object rotates really fast?

1. If an object rotates too fast, it will fly apart. An object must be dense enough to retain its rotation and not fly apart.

2. Some objects in the universe are so tightly packed that they can sustain a very high rate of rotation. Balls of neutrons, known as *neutron stars* or *pulsars*, are the densest state of matter known. A thimble-full of the material of a pulsar placed on a scale would balance with a herd of 50 million elephants.

3. These neutron stars have such high gravity that they can spin enormously fast without any danger of flying apart. Nothing has a chance of taking shape in this gravity, making neutron stars the most perfect spheres in the cosmos.

V. Finally, the observable universe is also a perfect sphere.

A. The universe was born 13 billion years ago. From any direction we look, the farthest we can see is 13 billion light years, because at that point, we see the beginning of the universe.

B. Our "visible edge" is 13 billion light years in every direction, and we are at the center of that horizon.

**Suggested Reading:**

Feynman, Richard P. *The Character of Physical Law*. Cambridge: MIT Press, 1973.

**Questions to Consider:**

1. Why is it more useful to ask why something is not round than why something is not flat?

2. In general, which are rounder, high-mass objects or low-mass objects? Why?

# Lecture One—Transcript
## On Being Round

Welcome to My Favorite Universe. I'm Neil deGrasse Tyson. I'm an astrophysicist and I'm Director of the Hayden Planetarium in New York City, part of the Rose Center for Earth and Space of the American Museum of Natural History. I'll present to you, not lectures that are part of some curriculum, but in fact, I've combed the universe for my favorite subjects, and I'm going to spend twelve lectures bringing those favorite subjects to you. These are subjects about which I've thought quite a bit. I've written about each one of them in the form of essays that have appeared in *Natural History* magazine. I've written many essays, and these are some of my twelve favorites, organized in four groups of three.

The first set of lectures is "On Being." "On Being," is a series of three lectures that presents to you some of the machinery of what shapes and forms the cosmos. Machinery that we explore and investigate and discover here on Earth's surface and find out that, in fact, the same properties of matter and forces and energy reveal themselves elsewhere in the cosmos, enabling science to be an enterprise et al.—the fact that, what happens on Earth happens elsewhere.

The next set of three is all about the "blood and gore" of the cosmos; it's "Cosmic Catastrophes" really. We'll learn about how black holes rip apart stars. In fact, we take a person and toss a person into a black hole—not really—but we describe what would happen if we performed that experiment, and that's pretty gory; but, it is the real universe and we need to know about that sort of thing. Also in those lectures, we talk about cosmic impacts and what role they played in the history of life on Earth.

We start at the beginning for the next set of three, "The Big Bang," and move forward and present to you the frontier of our understanding of the biggest questions of the cosmos: How did the universe get here? How has it evolved? How will the universe continue to evolve into the future?

We end on a set of three lectures, which broadly refer to "The Search for Life in the Cosmos." It just so happens, any time I'm on an airplane and someone finds out I'm an astrophysicist, I always get asked a series of questions and the first one, 99 times out of 100 is, "Is there life elsewhere in the cosmos?" We've saved the best for last. We've saved the most enticing subjects for last because everybody wants to know whether or not we're alone. We'll tackle those subjects and we'll do it broadly. We'll talk about how to find environments where you might expect to see life, and not only that; if we find life, what might that life might be like. Might it be like us or nothing at all like us?

Welcome to this series of lectures and let's begin. The "On Being" series is a set of three lectures: "On Being Round," "On Being Rarefied," and "On Being Dense." These are three sort of properties of matter that, by describing them, take me places in the cosmos and have me describing objects that will show up again in later lectures. It's not only a philosophical introduction; it's also an introduction, in some measure, to what objects are brought in to make the examples that we give.

Let's begin with "On Being Round." You've looked around at globes and pictures of orbs; why is so much stuff in the universe round? I don't know if you've ever asked that question, but now that I bring it to your attention, why not? Let's ask it. "Why is so much stuff in the universe round?" We learn that the forces that make things round operate on small scales as well as large scales, because it refers to the energy of a body, and energy likes descending to the lowest energy state it can. In other words, if you build a house of cards, the cards don't really want to stay built. If you jiggle it, they all fall down and they get to a lower point on Earth's surface than they once were. This is profound. It wasn't simply that it's a shaky thing; it's that it's natural for things to want to go to the lowest levels possible. That's a natural consequence of physical law.

Look around at all the round things in the world. Is there anything nature makes that's not round? Yes, there is. There are a few things. To name a couple, crystals are not round. Crystals have sharp edges. There is some action going on geo-chemically that makes crystals, and it's beautiful. Crystals are beautiful. That's an exception. Plus, you have fracture rocks. Those have angles to them. But, by and large, if you look around in the cosmos, there are very few things that make angles. Things are round, and in the best of the cases, they make beautiful spheres.

Let's make a short list of round things. We've got soap bubbles; soap bubbles are round. You had those as a kid, of course. Soap bubbles, moons, planets, stars, galaxy halos; even the observable universe is round. It's a perfect sphere centered on us. We'll end with that concept to learn exactly what we mean by the "observable universe," and why it's a perfect sphere centered on us.

The way this works is, we have forces that want to shape the object in such a way that the surface is minimized. What do we mean by that? Let's take a look at a soap bubble. Here we have a child blowing soap bubbles. We've all done this as children, and I wish more adults would do it into their later years. Soap bubbles are fascinating, because no matter the cavity through which you blow this soapy liquid, out the other side comes a sphere. What goes on there is—since a sphere is the shape that encloses the largest volume with the least surface—if a soap bubble wanted to be another shape, then the soap film would not be as strong as it is as a sphere, because in any other shape it has to give more of itself to cover that surface area, making it thinner in one place and not another. It would just break.

This generalized feature can be revealed if we take a look at the fact that, on a cube there are parts of the cube that are more distant from the cube's center than others. Every corner is farther from the center than the middle of the sides. In a way, that makes the sides kind of weaker. If this were an orb that had gravity, a corner, which is farther away, would be some kind of mountain; and the forces that take over, that enable gravity to have made the planet in the first place, will want to make that mountain smaller. A rock on the side of that mountain will roll down and fill up the low areas, and this process continues until you get something that much more closely resembles a sphere and not a cube.

Soap bubbles aren't the only things that make spheres. Raindrops make spheres. There's that classical depiction of a raindrop that looks like a little teardrop; but in fact, if you were to watch rain as it comes out of the clouds and it falls down, it is a perfect sphere coming down. What's holding it together? There's this thing called "surface tension." On the surface of liquids—the boundary between the liquid and the air—on that surface the molecules kind of grab onto each other. The act of establishing the surface of the liquid gives it a kind of tension that wraps the liquid; so that when it's left to fall on its own, it wraps itself into a perfect sphere, once again, making the most efficient shape it possibly can.

Have you ever wondered how you make ball bearings? These little round things. You can't lathe them. They're too small. One way to make them is, you have liquefied metal and you have an evacuated tube, and you take eyedroppers of this liquid metal and you drop them down the tube. As it comes down, it cools and hardens; but before it hardens, that drop of liquid metal turns into a sphere, it hardens, it lands down, and you collect your ball bearings at the bottom. That's one way you make ball bearings. Another way is, if you were weightless in the space shuttle, orbiting the Earth; you could just take your liquid metal, heat it and squirt it out of the eyedropper. The liquid comes out of the eyedropper, floats in front of your face, makes a sphere, and just hardens right there. In fact, you can make the most perfect ball bearings ever if you had a ball bearing factory in a weightless environment, like on the space station.

You're relying on the surface tension of the metal to bring itself into a sphere. By the way, the only metal that's like runny liquid at room temperature is, of course, mercury—not the planet, of course, but the metal. When I was a kid, we had little games where there was a mercury bead that would roll around and you would go through a maze; but now we've learned that mercury is bad for your health. It interferes with your development of intelligence as a child.

With these games you notice that the mercury has such high surface tension it pops into a sphere just sitting here on Earth. The surface tension was greater than that which would otherwise make it flatten out. Greater than the force of gravity that would otherwise make it look like a bead of water sitting on a nicely waxed car, that looks like it wants to pop into a sphere, but doesn't quite have enough surface tension to wrap itself up into its little package, it's little spherical package.

If a sphere minimizes how much surface area you have, you might ask the question, "Why isn't everything a sphere?" Why aren't goods offered for sale in a grocery store spheres? You could save billions of dollars on packaging. Imagine a box of Cheerios, and it's just a sphere. That would minimize how much box you would use. Or a can of beans. Make spherical cans of beans. That would be so cool. I thought of advising the industry on this, but, of course, you realize there's another problem. Spheres roll, so it would be very hard stacking these things on the shelves. It would just be a mess in the grocery store. There are other factors that contribute to this; but it doesn't change the fact that if you have a spherical container, you have maximized the volume and minimized the outer surface, both at the same time. Nature knows how to do that beautifully.

Let's ascend from Earth and think about the solar system and what kind of spheres we have going there. As I already alluded to, there's kind of a conspiracy of gravity and energy that gives you spherical shapes. The Sun is one example. The Sun, everyone's favorite star, is basically a perfect sphere. There's some structural form on the surface; but, if you step back from it, it's a perfect sphere. It's a big ball of gas.

Gas is easy to make into a sphere. Any part of the gas that is a little higher than the other part will sink down and push the other side up. It might oscillate for a bit, but it's going to end up forming a perfect sphere. Do you know why? Because it all wants to do is get as close to the center of gravity as possible. The only way it can do that is to make the high places low. That minimizes how much total energy is expressed in that field of gravity.

That's why rocks roll down hills instead of up hills. They roll down hills into the ditch and will stay there forever unless you go and pick it up and put it up somewhere else again. Things have the tendency to fill in low places. That's the Sun. It's nice and big, a big gaseous ball.

Let's go to something that's one-tenth the diameter of the Sun, the planet Saturn. Hold onto the ring for a moment; we'll get back to Saturn's ring. Look at that ball. It's a gorgeous spherical ball. It's one-tenth the diameter of the Sun. It's also a sphere. Look carefully. Actually, it's a little bit flattened, top to bottom. We'll get back to that, too, because we'll learn that if something's not a sphere, it tells us something about what's going on in the environment of that object. Why something is not a sphere is more interesting even than why it was a sphere in the first place, because it wants to be a sphere, and if it's not one, it's telling us something.

Let's go to an object one-tenth the size of Saturn: Earth, a beautiful, blue ball, looking like a marble in the sky in an image taken by Apollo 17. Let's go to one-fourth the size of Earth. We all know what's coming. One-fourth the size of Earth, we get the Moon, once again a sphere. Gravity has no problems turning a wide range of size of objects into a sphere.

You can ask the question, "Does gravity always win?" Is there a point where gravity fails in its attempt to turn something into a sphere? The answer is yes. Gravity does fail. There is a point where the object is so small, and the field of gravity is so weak, that— imagine you're looking at this mountain I described a moment ago and you nudge the rock; there's not enough gravity to have the rock roll all the way downhill. It just kind of goes a little lower, but that's about it. Otherwise, it's happy staying up there, because it doesn't weigh that much. There are such objects.

Let's go to something one-tenth the size of the Moon. What do we get? With one-tenth the size of our Moon, we get a moon of Mars, Phobos. Look at that poor thing. It has craters and it's got a funny shape. I wouldn't want to have been there when that got hit or punched or whatever. That's Phobos. You notice it's starting to not look like a sphere any more. It doesn't have enough gravity to have wrapped it into a sphere in the first place.

Let's go even smaller than Phobos. Gaspra, an asteroid that is one-tenth the size of Phobos, looks like a fine Idaho potato. It has rounded edges, sure; but it's not a sphere. Something's going on. It turns out, if the object is small enough, the chemical bonds of the stuff it's made of—which are no stronger in this case than they were for the Earth—are stronger than the force of gravity; and gravity is helpless in the presence of those chemical bonds. If you're born a rock you stay a rock. The action of gravity doesn't round them out. This is one asteroid, Gaspra.

I've got another one for you, about the same size. This is one of my favorites, called asteroid Cleopatra. This is a series of images of Cleopatra rotating. We call it the "dog bone asteroid." It looks like a dog bone; nothing like a sphere. Forces of gravity are helpless in their attempt to turn that into a sphere. By the way, it's also helpless in turning us into a sphere. I am not a sphere, the last time I looked. The chemical forces in my body enable me to retain my shape even in the presence of gravitational forces within me that might want to shape me into a sphere. I'm small enough so that I'm in full control.

Wait a minute now. The Earth picture that we saw is looking really spherical. We know that Earth has mountains and valleys and craters and cliffs, and there's the bottom of the ocean and it goes really deep and really high. It's got texture. It's got surface. It's the soul of geology going on in the surface of the planet; so what am I saying, that Earth is a nice perfect sphere? Surely it's got this structure on it. Let's analyze that structure. Let's think about that.

I come to my trusty globe, and we've got these mountain ranges. You can't feel it, but I can. I rub my fingers across it and I'm coming across Nepal, and there are some ridges there. That, of course, is the Himalayan Mountains. I come over to North America, and I feel the ridges, and up there go the Rocky Mountains, the Alaska Range up in Alaska. I come down to the Andes Mountains. The globe makers, in consort with the geologists, raised the surface of this globe in order to tell me where the mountains are.

Sure it would feel like that, you'll tell yourself. The answer is, no; it would not feel like that. Do you know why? Let's go to the deepest part of Earth's crust. Do you know where that is? Marianas Trench, far away in the Pacific Ocean, is 35,000 feet down. That's the deepest part of Earth's crust. That's five or six miles down. What's the highest point on Earth's crust? Mount Everest. How high up is that? It's 28,000–29,000 feet; that's five or six miles up. The total distance between the lowest part of Earth's surface and the highest part of Earth's surface is a dozen miles. That's the length of Manhattan Island. You can drive that in 12 minutes obeying the speed limit.

Earth—as much credit, as we'd like to give to the beautiful structures upon its surface—Earth, being 8000 miles in diameter, only has a 12-mile variation on its surface. It is as good a sphere as you could ever make in a laboratory. If Earth were shrunk down to the size of a cue ball, it would be the smoothest cue ball ever made. You would not even know where the mountains are, or the valleys or the hills. In fact, if I were just some cosmic giant and Earth was out there in the sky, and I took my finger and rubbed across it; I would not know—I would not be able to tell, what was water, what was land, what was mountain, what was valley—because the depth of the mountains and valleys would be the same as the depth of my fingerprint lines. It would go unnoticed.

Yes, gravity works. It works beautifully. We just happen to be really small and we can't make the observation we need to make that these fluctuations are only one six-hundredths (1/600) of the diameter of the globe itself.

In the rest of the universe we find some interesting things. We look around for places where objects are not spheres and ask, "why not?" We know nature wants it to be a sphere. Let's use the assumption that it wants to be a sphere and that we can learn a lot about why it's not.

Let's take, for example, tidal forces. Tidal forces really wreak havoc on spheres. What are tidal forces? It's simple. If you're an object, and there's a source of gravity over there somewhere that's tugging on you, the side of you closer to that source of gravity feels more gravity than the side of you that is farther away. It makes intuitive sense. The closer you are to the object, the stronger its gravity field. If I have some extent to me, this arm is going to feel your gravity more than this arm is and it might try to stretch it out.

That happens with the Moon. The Moon exerts tidal forces on the Earth. One side of the Earth is closer to the Moon than the other. The oceans respond to that, and the oceans on this side bulge out: high tide. The oceans on the other side are not pulled as much toward the Moon as these, and everything is sort of stretched in this direction and thinned in the perpendicular sides. You have two low tides, two high tides. That's the tidal force of the Moon acting on the Earth, turning it slightly out of sphericity.

This also happens in binary stars where you have two stars that orbit around each other. The Sun is only one star. We look around the cosmos and there are plenty of places where there are two stars in orbit around each other. If one of them happens to be a black hole—we have such cases—and another one is a blue supergiant or a red supergiant star—you'll learn about those in another lecture—it can become so large, that some of its material will be funneled down toward the black hole itself. As it fills up, the side closest to the black hole feels an extra tug and the shape of the giant gets distorted, and it comes to a kind of a point as it funnels down onto the black hole. I like to think that shape resembles a Hershey Kiss. It's a distortion from what would otherwise be a native sphere. Once again, you have gravity—tidal forces of gravity—changing the shape of things.

If one object comes too close to another object, tidal forces can rip the first object apart. That happened with Saturn's ring. Some object, an asteroid—a loosely bound asteroid or comet or some combination—came too close to Saturn, with one side of it closer to Saturn; and Saturn's gravity ripped it apart and scattered the whole thing into a ring. Eventually, those particles will fall out of orbit and Saturn will lose its ring. I'm happy to be around at a time where my favorite planet has the most beautiful ring in the solar system.

Another thing that can affect shape is rotation. Jumbo gas clouds are what give birth to stars. Anything up there has some kind of movement; but, as it begins to collapse, there's something called the *conservation of angular momentum*—a physics term. All it refers to is the fact that, if an object is big and is rotating slowly, as it gets smaller, in order to compensate for having gotten smaller, it speeds up. It's got to speed up. We've seen this before. You know this happens. It happens every day on an ice-skating rink. There's some ice skater who is turning like this, with her arms outstretched; and what do they do? They bring in their arms and what happens? They speed up. Let's try that. We can go slowly and then speed up. They could be much more effective at that if they carried weights in their hands, bringing more weight into their chest than just the weight of their fists.

That happens to gas clouds as well as happening on ice-skating rinks. As the gas cloud collapses, it begins to spin up; and there is a resistance for material to flow inward in the plane of rotation. It's like that centrifugal force you're trying to resist, that force of the rotating platter at the amusement park. If you're in the plane of the platter, it's kind of hard to work in because it's trying to throw you out to the outer wall. We call it a centrifugal force, but it's not really a force at all, just your tendency to fly off. Fortunately, there's support back there so that you don't fly off the amusement park ride; but, it's real laws, real forces of physics. This rotation preserves the plane, but continues to collapse from top to bottom. It has the effect of flattening a system.

The gas clouds make stars; they also make galaxies, galaxies with 100 billion stars in them. Let's make a galaxy. Here we go. Here's a big gas cloud and I'm going to make some stars early, so I make some stars. Now they have orbits that sort of come around like this, and that's fine. The rest of the gas is still collapsing and it meets in the plane of rotation. It sort of kept its shape out here, because that's where it met the plane of rotation. The rest was able to collapse down and we've got a flattened system. There are galaxies that look like this. We've got an edge-on spiral galaxy. Look at that. That's a flattened system seen edge on.

Some stars retain their spherical distribution, that's the bulge in the middle. The fact that this system is rapidly rotating meant it's not going to collapse in the plane, but it did collapse top to bottom. We've lost what was previously a full sphere and it went down to something flat as a pancake, flatter than a pancake, in fact. The Milky Way looks like this. It is as flat as a crepe. It's about a thousand to one (1000:1), diameter to thickness—a crepe.

We've got stars above and below the plane of the Milky Way galaxy, formed before it had collapsed. In fact, those stars are the skeleton of the sphere that was once there. I have a galaxy that has a much bigger bulge than our galaxy. It's called a "sombrero" galaxy. Look at that. It's beautiful—one of my favorite ones in the cosmos. About half the stars in this one formed before the thing collapsed and retained their spherical orbits, and the other half of the stars collapsed, preserved the plane of their rotation, and made the disk. This one is a hybrid of the collapsed story.

There is also general flattening. Simply, once you're made—if you keep rotating fast—you just get a little flatter. Earth is slightly bigger at its equator than it is at the poles. The circumference is 25,000 miles and we make one rotation a day. If you're on the equator, you're moving 25,000 miles in 24 hours. How fast are you going? You're going about 1000 miles an hour if you're just standing on the equator, minding your own business. That puts a stress on the ball, and so it will bulge out a little bit; so Earth has a bulge, an equatorial bulge. You might read about that.

Not only that, Saturn has an equatorial bulge. In that picture of Saturn we saw, Saturn is actually 10 percent shorter, pole-to-pole, than it is across its diameter—across its equator. Saturn also rotates much faster than Earth and is much bigger. In fact, if you're standing on the equator of Saturn, you're moving 22,000 miles an hour. That's going at such a good clip that Saturn has flattened for having rotated so fast. It becomes oblate. We have two words: "oblate" means you're flattened, and "prolate" means you're sort of elongated. I just think of hamburgers as oblate and hot dogs as prolate, if you need some examples. Especially when I'm hungry I think that way.

What happens when you rotate really, really, really, really fast? If you spin up Saturn even faster, you know what will happen eventually? Pieces of it will fly off. Imagine doing this experiment with the Pillsbury Doughboy. That's it. You take him and spin him; his arms fly off—end up on the walls. You've got to be tough enough to sustain that rotation rate and to stay in one piece.

There are objects in the universe that are so tightly packed and so dense that they can sustain a high rotation rate. They're balls of neutrons. We call them, sensibly, *neutron stars*. If they rotate and they pulse at us, they're also called *pulsars*; but they are neutron stars and they are densely packed. It is the densest state of matter known. Take the Sun; compress it down into a ball about 10 miles across. That's the density of a neutron star. Let's see how dense it is: scoop out a thimble-full of the material and put it on one of these balance scales. Do you know what you have to put on the other side of that scale in order to make it balance? A herd of 50 million elephants and that would balance one thimble-full of neutron star material. Equivalently, take the 50 million elephants, cram them into a thimble, and that's the density of a neutron star.

Neutron stars are extraordinary things. It turns out they have such high gravity that they can spin enormously fast without any risk of flying apart like the Pillsbury Doughboy. In fact, we look in the universe and we find these pulsars just spinning 30 times a second. When we first discovered these we said, "we don't know any form of matter that can hold together and spin at that speed." That's when someone was able to say, "Wait a minute. This is not ordinary matter. This is nucleonic matter." This is a dense ball of neutrons—and thus was the discovery of stars made completely of nuclear matter, neutron stars, otherwise known as pulsars.

Let me tell you how serious gravity is on them. We have mountain ranges here where rocks roll down the mountain. Go to a neutron star. I put a sheet of paper on a neutron star. You say, "What's the thickness of that sheet of paper?" The gravity is so strong that my effort to get on top of that sheet of paper—lifting myself to that thickness—will require as much energy as a rock climber on Earth ascending a three thousand mile cliff. That's some serious gravity going on, on the surface of a neutron star; and in fact, nothing has a chance of taking shape against the will of its gravitational force. Neutron stars are the most perfect spheres in the entire cosmos, nary a mountain range to be found.

My last point here is just the observable universe. It's simple. Light takes time to get from one point to another, and the universe has a birthday: thirteen billion years ago. Every direction we look, the farthest we can see is thirteen billion light years, thirteen billion light years no matter which way, because at that point we see the beginning of the universe. Our "visible edge" is exactly thirteen billion years in every direction; and we're at the center of that horizon. Just the way a ship at sea looks around and it's the same distance from every point on its own horizon. Wherever it is, it kind of takes its horizon with it. Just as is true here in the entire cosmos.

Let me just say that spheres are everywhere. They make a very useful tool. I'll end with one of my favorite jokes: How do you make a cow produce more milk? Ask that of someone who is an animal husbandry expert, they might say, "Well, change the diet of the cow, feed it better food." Whatever. Ask that same question of an engineer, they say, "Well, fix the machine. Change the suction rate." They'll try to change the machinery of the process. But ask that question of an astrophysicist, who lives in the world of spheres, where anything that is not in a sphere is better off thought of as a sphere and see how you adjust to make the real case. We would say, "Imagine a spherical cow."

# Lecture Two
# On Being Rarefied

**Scope:**

In this lecture, we look at rarefied phenomenon in the cosmos. In astrophysics, we use the term *rarefied* to mean "low density." We sometimes hear that a magician pulled a rabbit out of "thin air," but how thin is air, and are other components of the universe even thinner, or more rarefied, than air? This lecture examines those questions.

## Outline

**I.** We know that air is made of nitrogen and oxygen, but how dense is it? How many molecules of air would fit, for example, in a thimble, or about a cubic centimeter?

   **A.** The answer is about a quintillion—about the same number of molecules of air would fit in a thimble as there are grains of sand on an average beach. Air, then, is not really thin, if we are counting molecules.

   **B.** This quintillion particles of air in a thimble, has a certain weight that we call *sea-level air pressure*.

   **1.** *Pressure* is defined as "the force per unit area." Think of it as a weight.

   **2.** Sea-level air pressure is 15 pounds per square inch.

   **3.** Think of a square inch of space on the ground. From that inch, imagine cutting out a one-inch square column of air that continues all the way up through Earth's atmosphere. If we put that column of air on a scale, it would weigh 15 pounds.

   **4.** Air pressure is the weight of that column of air.

   **C.** If we put a suction cup over the square inch of ground, we are removing the air that was inside the pressure column and that was balancing the air all around it.

   **1.** Once we remove the air, the full weight of the 15 pounds per square inch is resting on the suction cup, and we can't pick up the suction cup because the atmosphere is pressing down on it.

   **2.** How much force do we need to lift the cup? The answer depends on the surface area of the suction cup. If it is 10 square inches, then we need a force of 10 x 15 pounds per square inch, or 150 pounds of force.

3. When we lift up the suction cup, the air immediately flows back in to fill the vacuum that we originally created.

D. If we travel to Mauna Kea in Hawaii, we are at 14,000 feet above sea level, and the air pressure is no longer 15 pounds per square inch. Because we're higher up, the column of air above us is shorter, and the pressure is about 10 pounds per square inch.

E. Earth's atmosphere extends for thousands of miles.

1. The boundary of our atmosphere can be defined as the place where the density of air can no longer be distinguished from the density of the space between the planets.

2. Most of the air in our atmosphere, though, is compressed down to the lowest levels. In fact, 99 percent of air molecules in our atmosphere are found below an altitude of 50 miles.

3. The atmosphere has some air molecules above that, but the air there is very low density compared with sea level or even with the density on top of a mountain.

II. Conditions above an altitude of 50 miles are very different from those on Earth. At altitudes of 50 to 100 miles, molecules collide less frequently, and the whole dynamic of their behavior changes.

A. A constant stream of charged particles, called the *solar wind*, travels through interplanetary space and comes near Earth. Because these particles are charged, they respond to the charges in the magnetic field of Earth. Positive charges go to one pole; negative charges, to the other pole; and they spiral down toward Earth's magnetic pole.

B. As these particles travel, they start to collide into Earth's atmospheric molecules at altitudes of 50 to 100 miles, and they never make it farther down. As low as the density is at those altitudes, it is still high enough for these particles to hit air molecules.

C. When the particles of the solar wind hit air molecules, the molecules become "excited" and release light—photons of energy in the form of blue, yellow, and green light. The result is an *aurora*, a display of light in the sky.

III. Another material in the cosmos that is thin is the *solar corona*, the crown of the Sun.

A. The solar corona can be seen only during a total solar eclipse. When the Moon moves in front of the Sun, the light of the Sun is removed, and we can see its glowing outer atmosphere, which is not bright enough to reveal itself when the Sun is visible.

B. What is the solar corona? For a long time, scientists believed that it was made up of glowing gas.

1. When the light from the corona was passed through a prism, however, it was broken up and its component colors and elements could be studied.

2. When the light of the corona was studied in this way, scientists found the signature of an unknown element, which was called *coronium*.

3. Later, scientists learned that under the low-density conditions in the outer atmosphere of the Sun and at very high temperatures, iron emits a signature of light. Highly ionized iron—that is, iron that has lost most of its electrons—has an unmistakable spectral signature, but the conditions that would make iron behave in this fashion had never existed on Earth.

IV. Let's move once again out into the galaxy to the asteroid belt, which many of us think of as a dangerous shooting gallery of asteroids.

A. If we compressed the asteroid belt, however, we would find that its mass is, in reality, only two or three percent of the mass of the Moon. In addition, 75 percent of that mass is contained in four asteroids, and the rest is scattered for 100 million miles in a one and one half billion-mile orbit around the Sun.

B. The asteroid belt has a much lower density and is much less dangerous than what we might think. In fact, four of our spacecraft, *Pioneer 10* and *11* and *Voyager 1* and *2*, went through the asteroid belt without incident.

**V.** Let's move again, farther out, into interplanetary space.

    **A.** The density of matter in interplanetary space is 10 molecules per cubic centimeter. Remember that the density of air on Earth at sea level is a quintillion molecules per cubic centimeter. By comparison, interplanetary space is about as good a vacuum as can be created.

    **B.** In the vacuum of interplanetary space, we find other objects that are also visible but quite thin, such as comet tails.

        **1.** The tail of Halley's comet, for example, is 100 million miles long, yet it is quite thin. If we collapsed the whole tail to atmospheric density, it would fill a cube that was about one half mile per side.

        **2.** What are comet tails made of that makes them so visible despite their thinness? Comet tails reflect some light from the Sun and emit some light of their own for having been excited by the high-energy photons of the Sun.

        **3.** A comet tail is a visible stream that has 1000 times the density of interplanetary space, but that is still low density compared with the atmosphere of Earth.

        **4.** Spectroscopy has revealed that comet tails contain cyanogen (CN), a deadly poison. But the density of cyanogen in a comet tail is so low that Earth can pass through the tail, and the poison has no effect on life.

**VI.** What is the density of the Sun? The Sun is very dense at its core and less so at the surface.

    **A.** The average density of the Sun is about the density of water and of humans. The density of water is one gram per cubic centimeter; the average density of the Sun is 1.4 grams per cubic centimeter. A scoop of the average material of the Sun would sink in the bathtub but not that quickly.

    **B.** In five billion years, the Sun will be about to die. It will have swollen up and become a red giant. The Sun will be 100 times bigger in diameter than it is now, but no mass will have been added to it, so its average density will drop to about 1/10,000,000 of its current density.

        **1.** The surface of the Sun will be very close to Earth, which means that our atmosphere will evaporate, the oceans will boil and evaporate, and life will be vaporized.

2. As Earth orbits the Sun, as rarefied as the Sun's material is, Earth will plow through some of that material.
3. That material will resist the motion of Earth, and Earth will lose its orbital energy and spiral down into the center of the Sun.
4. As rarefied as the Sun becomes as a red giant, it will still impede the motion of Earth.

VII. Interstellar space, which would take about 25,000 years to reach, is even less dense.
   A. The density of interstellar space is about 1/10 of what it is in interplanetary space—a couple of atoms for every few cubic centimeters.
   B. If gas clouds in interstellar space are near a star, they are rendered visible. The star reflects light off the gas cloud, and the molecules of the cloud become excited and release light.
      1. Again, another unknown element with a distinct spectral signature was discovered in these gas clouds.
      2. This element, nebulium, turned out to be oxygen in a peculiar state of temperature and very low density.

VIII. If interstellar space is empty, intergalactic space is even emptier.
   A. Intergalactic space has no dust, no comets, no stars, no moons, and no planets, and its density is about one atom per cubic meter.
   B. A 200,000-kilometer cube of interstellar space has about the same number of atoms in it that are in the air in your refrigerator.
   C. Most of the universe is this kind of vacuum.

IX. We know that the universe is expanding and that there is not enough mass in the universe to exhibit enough gravity to halt that expansion.
   A. How much mass, or density of matter, would it take to balance the expansion of the cosmos?
   B. The answer is only about 10 particles per cubic centimeter.

**X.** Suppose we could create the perfect vacuum, in which there were no particles. What would we measure?

    **A.** The relatively new field of quantum mechanics describes nature on its smallest scale, at the level of atoms. At this level, quantum mechanics finds new forces of physics and new behaviors of matter.

    **B.** Quantum mechanics predicts that the perfect vacuum couldn't really be a vacuum. A vacuum is seething with *virtual particles.*

        **1.** Virtual particles are matter/antimatter pairs that pop into and out of existence in such a short period of time that their existence can't be measured.

        **2.** This theory sounds like science fiction, but few have questioned it because quantum mechanics has been correct in its other predictions.

        **3.** These particles popping into and out of existence in a vacuum create a pressure in the environment. Pressure is the density of energy that has the opposite effect of gravity; instead of bringing things together, it pushes them apart. This pressure is called the *vacuum energy.*

        **4.** If we try to calculate how much vacuum energy is in the universe, we get a number that doesn't make sense; we know that something is wrong with our calculations. But we do know that a pressure exists in the cosmos that will never let the galaxy collapse into itself.

    **C.** A few years ago, scientists also discovered that the universe has an anti-gravity pressure operating on it, called *dark energy.*

        **1.** We don't know what it is made of or where it came from.

        **2.** Vacuum energy, however, could explain the existence of dark energy.

        **3.** This is the very frontier of our understanding.

**XI.** As a final question, we might ask, "Is there a limit to nothingness? Is there a place outside of space with even less in it?"

**Suggested Reading:**

Tyson, Neil deGrasse, Charles Liu, and Robert Irion. *One Universe: At Home in the Cosmos*. Washington, DC: Joseph Henry Press, 2000.

**Questions to Consider:**

1. How do our best laboratory vacuums compare with the vacuum of interplanetary space?

2. Why is aurora formed so high up in Earth's atmosphere?

# Lecture Two—Transcript
## On Being Rarefied

Welcome back to My Favorite Universe. Today I want to talk about one of my favorite subjects, and that is rarefied phenomenon—rarified objects in the cosmos. There is a lot out there that is dense; but there is so much more that is rarefied and that it is certainly deserving of its own presentation.

Let's begin on Earth. Let's begin with what's most familiar. We've all been to a magic show and seen a magician with a top hat pull a rabbit out of a hat. That's usually described as pulling a rabbit out of "thin air." Air is something we think of as very low density, very rarefied. In fact, if someone didn't tell you there is air between us, you might not think there was anything there. You know there's air because wind blows trees and you see storm systems. You see all this; but, if you just plopped here on Earth, there's no obvious measure that there's air there, that there's anything there at all—nothing blocking your view.

Let's ask ourselves, what is this air made of? We know it has nitrogen and oxygen, sure. How dense is it? Let's look at how many nitrogen and oxygen molecules of our atmosphere at sea-level air pressure, would fit into a thimble, a cubic centimeter of volume. That has about the same number of molecules as the number of grains of sand on an average beach: about a quintillion, one with 18 zeros. There's nothing thin about air at all, if you're counting molecules.

We still like thinking of air as "nothing" or something close to nothing. That concept goes very far back. Look at the famous Aristotelian elements: Earth, air, fire and water. It's one of the base elements from ancient thinking. There was a fifth element that was lighter than all four of those. A fifth element, a "quint essence." *Quintessence* it was called, and that was the substance of the cosmos—space. That rose above the Earth and above the air, because it was even lighter than air. It was not where Earth is. It was in the heavens, filling the cosmos.

Let's think a little more about these quintillion particles per cubic centimeter. There's a certain weight that that has. We call it *sea-level air pressure*; but pressure is a "force per unit area," so let's think of it as a weight. Let's take a square inch. Sea-level air pressure is about 15 pounds per square inch. Now what does that mean? It means something simple. Put a square inch on the ground and cut a column of air through the atmosphere one square inch in cross section. Start from the surface of the Earth and go all the way out to the farthest reaches of the atmosphere. That is a column of air. It goes up hundreds of miles, even thousands of miles. Take that, put it on a scale and weigh it. It is 15 pounds. That's what air pressure means. It is the weight of the column of air above.

What happens when you have a suction cup? If you take a suction cup and you press it down onto the ground, you're removing the air that used to be inside that suction cup balancing the air all around it. Once you take out the air, the full weight of the 15 pounds per square inch is resting on the outside of that rubber. If you try to pick up that suction cup, you say, "Oh, it's sucking me down." No, the atmosphere is pressing down on the surface of that suction cup, preventing you from pulling it up. How much force do you need to lift it? What's the surface area of the suction cup? If it's 10 square inches, then you need 10 x 15 pounds per square inch. You need 150 pounds to pull it. If you could put 150 pounds on it, you could pop the thing right off the ground. That's how suction cups work. The moment you lift it up, air swiftly swoops in to fill the vacuum that had been left there by pressing the air out in the first place.

There's an old adage, "Nature abhors a vacuum." Wherever there is a vacuum, nature collapses down on it to get rid of the vacuum. We have this idea that somehow a vacuum is rare, or uncommon, or something that nature does not like. I'm an astrophysicist, and my concept of nature of is not just what happens on Earth's surface; it's what happens in the cosmos. In the cosmos, in fact, nature loves a vacuum. We'll see more about that in just a few moments.

Let's keep playing with this notion of atmospheric pressure. If you go to a mountaintop—let's go to one of the highest mountaintops where there are large telescopes, Mauna Kea, Hawaii. In Hawaii, that is at 14,000 feet. It's not 15 pounds per square inch any more. You're higher up. The column of air above you is shorter. There is less air; there is less pressure. It's about 10 pounds per square inch. There is less air. It also means there is less oxygen to breathe. So, at the console for all of the telescopes, there is an oxygen tank in case you start getting giddy. In case your data taking starts becoming compromised, you can whip out the oxygen and that will help you.

Earth's atmosphere, as I said a moment ago, extends for thousands of miles. We don't think of it that way because we have the space shuttle and things orbiting relatively low compared with that, and we say that they're out in space. In fact, the boundary of Earth's atmosphere can be defined by where the density of air is no longer distinguished from the density of space between the planets. At that point you say, "okay, Earth's air is here and the rest of the solar system is there."

That being said, we know most of that air is compressed down at the lowest level. In fact, 99 percent of all the air molecules on Earth are found below a height—below an altitude of 50 miles. There's still a little bit of molecules up above that, not many. Conditions are wholly unlike what you find in the laboratory; very low density compared with sea level—compared with Mauna Kei, very low density.

At those altitudes—50 to 100 miles up—molecules collide less frequently, so the whole dynamic among the molecules has changed. Here's an interesting fact: there's a stream of particles coming from the Sun. They're called the *solar wind*, a stream of charged particles. This is well known ever since we've been studying the Sun. These charged particles travel through interplanetary space and they come near Earth. Since they have charge, they respond to the magnetic field of Earth. Positive charges go to one pole, negative go to the other, and they spiral down in toward Earth's magnetic pole.

It is not a free trip. There is atmosphere there, so they start colliding into atmospheric molecules. You know when that begins to happen—between 50 and 100 miles up. They never make it low down. As low density as that is between 50 and 100 miles up, it's a high enough density for these things to hit. When they hit, they "excite" the molecules. Molecules love getting excited, because you excite a molecule and upon being excited, it says, "Okay, how will I de-excite myself?" One of the ways is that it releases light, photons of energy—green light, yellow light, blue light—and, as it releases this energy, it creates a display of lights in the sky, better known as the *aurora*.

One of my favorite images of the aurora is not one taken from Alaska, although some of the best terrestrial images come from there. My favorite is taken from the space shuttle. If we take a look at the image now, we see the tail end of the space shuttle. The space shuttle orbits a couple of hundred miles above Earth's surface. You notice in this image that, of course, the aurora is below—is lower in altitude than the space shuttle. Just look at the light emitted. That is the emitted light of molecules of Earth's atmosphere having been excited by particles streaming from the Sun. This only happens in very low-density environments, such as what happens in the upper atmosphere. You get this in the Northern Hemisphere as well as in the Southern Hemisphere, and you might be more familiar with them as the aurora borealis and the aurora australis.

Another place where material is thin is the Sun, the *solar corona*. You hardly ever get to see the corona unless you're eyewitness to a total solar eclipse. I've seen one in my life, unforgettable, that was off the coast of Northwest Africa some years ago. What happens in a solar eclipse is the Moon moves in front of the Sun, eclipsing the Sun—occulting the disk of the Sun. When that happens, when the light of the Sun is removed, you see the glowing outer atmosphere of the Sun, which is not bright enough to reveal itself when the Sun is visible. You have to blot out the disk of the Sun and then you see the solar corona, the crown of the Sun.

We didn't know much about the solar corona for a long time. What is it? It's just some glowing gas. That's what we assumed, but we didn't know much about it. Then we studied the light. We took the light and passed it through a prism and studied the features of the light, the component colors. When you pass light through a prism, it breaks up that light into its component colors, just the way sunlight moves through raindrops and makes a rainbow—red, orange, yellow, green, blue, violet. Sunlight is composed of those colors.

If you do the same thing for the solar corona, you can study what kinds of elements are present in that corona. We found a signature of an element nobody had ever seen before. We invented a name for it. We called it *coronium*, because we didn't know what it was. There was no laboratory counterpart for it. We didn't know where it might fit in the periodic table of elements. We didn't know anything, but we had this placeholder name for it.

It turned out, under the low-density conditions in the outer atmosphere of the Sun, under very high temperatures, millions of degrees; iron emits a signature of light, highly ionized iron. This is iron that has lost most of its electrons, heavily ionized. It has a spectral signature, unmistakable, but we had never ionized iron that much before. We had never subjected it to low-density environments such as the corona before. We had no way to understand it until we figured out that it was under very low density and very high temperature.

The solar corona is an example of something that is quite visible and quite beautiful, but is also very, very low density in the cosmos. By the way, if it were much higher density it might be more visible, even with the Sun in the picture. You really have to blot out the Sun in order to see the corona. The temperature of the corona is millions of degrees. Millions. That's how hot it had to be to ionize iron as much as it got ionized. The surface of the Sun is a "mere" 6000 degrees. It remains a mystery how you can get something that hot sitting above something that cold. That is a major frontier of solar physics. We have ideas, but no true consensus just yet.

Let's keep rising up. How about the asteroid belt? We all have our favorite vision of the asteroid belt as this shooting gallery. Don't dare take your spaceship through the asteroid belt. They'll come and knock out your antenna. It will be really dangerous. Wait a minute. Let's look at the facts.

Take all the mass of the asteroid belt and combine it together and weigh it. How much is it? What do you get? You get about two or three percent of the mass of the Moon, of our Moon. No, it's not some jumbo planet that got shattered and all its debris is still there. It's like 2 ½ percent of the Moon. Of all that mass, put 75 percent of it in only four asteroids and take the rest and scatter it for 100 million miles wide in an orbit around the Sun—1½ billion miles around. That's the asteroid belt.

It is still more dangerous to go through the asteroid belt than to not go through the asteroid belt, but it is much less dangerous than most people ever think of it to be. There is much lower density of material than we ever credited it for having. Our four spacecraft that have gone to the outer solar system and have left beyond the orbit of Pluto—that's *Pioneer 10* and *11* and *Voyager 1* and *2*—those four spacecraft traveled through the asteroid belt without incident, just as we suspected.

Let's keep going out into space. We learned what the density of matter was on Earth. At sea level, it was a quintillion molecules per cubic centimeter. Out in interplanetary space, we're down to 10 per cubic centimeter. Now there's a vacuum for you. That's about the best we can get with laboratory vacuums. Here's a sketch of one. In the compartment that has a little circular part—inside that cavity is where they're pumping to remove as many molecules of air as they can to have as low a density environment as possible. Then you can do interesting experiments, particularly with the excitation of molecules.

In the vacuum of interplanetary space, there are other objects that are quite visible, but are also quite thin: comet tails. Comet tails are long—big comet tails. A good comet, like comet Halley, has a 100-million-mile-long tail. In the last several years, comet Hale-Bopp and comet Hyakutake were great comets, great in the classical sense of the word "great." You get a few of those in a lifetime, so make sure you put in your schedule to go see them, as many of us did.

What of the comet tail? As visible as they are, what are they made of? They are still thin, but they're remarkably visible for their thinness. They reflect sunlight and they emit light of their own for having been excited by the high-energy photons from the Sun. It's a visible stream, but not very high density. Yes, it's denser than interplanetary space, but not by much. The comet pioneer, Fred Whipple, once described comet tails as "the most ever made of the least," 1000 times the density of interplanetary space, which is still small compared with ordinary Earth air.

Let's quantify how thin the comet tail is, even though its density is high compared with interplanetary space. Let's take the whole tail, collapse it to atmospheric density and ask, "How much do you have?" That's a good way to compare. Take the 100-million-mile-long tail, and let's make atmospheric pressure out of that. You could fill a cube maybe a half mile on a side. That's all it is. Fred Whipple was right; it's the most ever made with the least.

When people first invented tools of spectroscopy, studying the light from objects and looking at what material was in them—what spectral signatures permeated the analyzed light—they found that comet tails had cyanogens (CN), a deadly poison. It would kill you, posthaste. When comet Halley came around in 1910, the public didn't know about how low the density of the comet tail was and how it was barely anything. No matter what you found, it was not enough to do anything to anybody. What scared people, was that we knew that Earth was passing through the tail of Halley's comet. It passed through the tail—it actually did—intersecting some of that gas. People got all worried. Charlatans were selling anti-comet pills, making a lot of money on people's fears. If you knew a little bit about how thin it was, you would have saved your money. So knowing about rarefied phenomenon in the cosmos can be good for your pocketbook, on occasion.

That takes us to something not widely appreciated. Let's look at the density of the Sun. It's gaseous, but you might think it's really dense. It is very dense in the core, less dense on the surface; but, if you averaged it, it's about the density of water. So, too, are human beings. That's kind of interesting. Water has a density of one gram per cubic centimeter. Human beings, we basically float or sink or somewhere in between, so that puts us at about one gram per cubic centimeter. The Sun is actually a little heavier. The Sun is just a little heavier at 1.4 grams per cubic centimeter. If you took a scoop of the average material of the Sun, it would sink in your bathtub. But, it would not sink that fast.

What will happen in five billion years? The Sun will be about to die in five billion years. It will have swollen up and become a red giant. The sun will become 100 times bigger in diameter than it is now. Of course, no mass will have been added to it, so the average density will drop. It will drop to one millionth of its current density. By then the surface of the Sun will be very close to Earth. That will be a bad, bad day on Earth. The atmosphere will evaporate; the oceans will come to a rolling boil; and they, too, will evaporate. Life will vaporize.

Ignoring those complications, we find that, as rarefied as the material of the Sun is, as Earth orbits the Sun, it will plow through some of that material. That material will resist the motion of Earth, and Earth will end up losing its orbital energy and spiraling into oblivion, down into the center of the Sun. That's going to happen in five billion years. As rarefied as the Sun will become as a red giant, it will still impede the motion of the Earth.

Let's go to interstellar space. We go beyond the planets, beyond the Oort Cloud of comets that surrounds the solar system. It's even less dense in interplanetary space. The nearest stars are far. For our space probes, *Pioneer 10* and *11* and *Voyager 1* and *2*, at those speeds, it will take 25,000 years to get to the nearest star. It is far. It is far, it is big and it is empty. There you'd be hard pressed to find a couple of atoms—a couple of atoms—for every few cubic centimeters. That's one-tenth of what it is between the planets, much better than most vacuums ever created on Earth.

Gas clouds that live between the stars are rendered visible—much like comets that come near the Sun—gas clouds in the galaxy are rendered visible if they're near a star. The star reflects light off of them and they get excited and release light from their own atoms and molecules. We have special telescopes to detect this.

In these gas clouds there was another spectral feature that no one knew about, and no one had ever seen before. In the nebulae of space we found an element. We didn't know what to call it; we called it "nebulium." Have you ever heard of nebulium? What did nebulium turn out to be? It turned out to be oxygen in a peculiar state of temperature and density. What kind of density? Very low density. The kind of densities that was very uncommon on Earth and very hard to create in the laboratory, although now we can do it routinely.

Interstellar space has got stars. It's got gas clouds. But, it's pretty empty between those places. However empty that is, it's emptier between the galaxies. That's empty between the galaxies. That's empty. There's no dust, there's no stars, there's no planets, there's no moons, there's no comets, there's no comet tail. It's empty. It's as empty as you could possibly find. There, you get one atom in a cubic meter—a cubic meter. Another way to think of it is, take a 200,000-kilometer cube. How many atoms are in that 200,000-kilometer cube? There are about the same number of atoms as in the air in your refrigerator. That's empty. That means that most of the entire universe is a vacuum. As I said earlier, the cosmos loves a vacuum.

You might have heard that the universe is expanding. There's not enough mass in the universe to exhibit enough gravity to halt that expansion and bring it back. We can ask the now academic question, "How much mass would it take to balance the expansion of the cosmos?" How much density of matter is required for that? It's got to be more than the one particle per cubic meter. How much more? The answer is 10 particles per cubic meter. If the whole universe had 10 particles per cubic meter, that would be enough to halt the expansion and send us right back from whence we came. That doesn't sound like much. It's not that much. Just more than what it is now.

What again of this notion of perfect vacuum? Suppose there were no particles, does that even have any meaning? How do you do that? Does such a place exist? We see from our favorite vacuum pump, once again, that we're making great strides to try to create the vacuum of interstellar space, intergalactic space. It may be impossible, but let's imagine the day comes when we can create the perfect vacuum—like the perfect storm—the perfect vacuum inside of this cavity. What would we measure? It turns out, the field of quantum mechanics, a very successful branch of physics discovered in the 1920s, describes nature on its smallest scale. It tells us what atoms and molecules are doing, which is wholly unfamiliar to the kinds of things we see happening in every day life. All kinds of new forces take effect, forces of quantum mechanics, new forces of physics, new behavior of material.

Quantum mechanics predicts that in the vacuum, the vacuum can't really be a vacuum. It is seething with these things called *virtual particles*, which are particle pairs—matter and anti-matter pairs—that pop into existence, annihilate, and then go out of existence again. They live too short a period of time to measure their existence. This just sounds like a cartoon. It sounds like science fiction or not like science fiction; it sounds just like fiction. So much else that quantum mechanics has predicted has come true that no one has the audacity to assume that this prediction is somehow going to be false when every other one has turned out to correct.

What does that mean, if particles are popping in and out of existence in the vacuum? You can measure what effect this has on the environment, and it creates a pressure, a density of energy that has an opposite effect of gravity. Instead of bringing things together, it pushes things apart. We call this the *vacuum energy*. You know something? If you calculate how much vacuum energy there is in the entire cosmos, you get a number that is kind of insane. Something is wrong with our calculation; but we do calculate that, in fact, there is a pressure on the cosmos that will never let us re-collapse, no matter how much matter we would ever find.

Recently we discovered that the universe has a kind of anti-gravity pressure operating on it. It's a recent discovery a few years ago. This is this famous *dark energy*. Nobody knows where it came from.

Nobody knows what it's made of, but we know it's there. This vacuum energy, from these seething virtual particles predicted by quantum mechanics, could explain that; except, that when you calculate it, the numbers are wildly different from each other. We don't know what the problem is. It remains a frontier of our understanding of the cosmos.

We feel a little better that there is something that we know about that can create this negative pressure on the universe, forcing it to expand exponentially into the future. Even though the calculation doesn't work out, it's far too much compared with what we see. It's encouraging, because before we had these ideas, there was nothing we could conceive of that could resist the titanic forces of gravity on a cosmic scale.

Is there some limit to this nothingness? I talk about "empty space," but I still use the word "space." I'm still saying there's a thing called "space." Suppose there is a place where there wasn't even space? What is that? Maybe that's outside of our universe—if there is such a concept—because, if in space there is nothing, then perhaps outside of our universe, where there is no space, there's not even nothing. What shall we call that? We would call it "nothing nothing."

# Lecture Three
## On Being Dense

**Scope:**

As a follow-up to our last discussion, this lecture examines what it means to have density. We will talk about different kinds of density, objects in the universe that are extremely dense, and some mysteries of density. We conclude by noting the usefulness of an understanding of density as a tool for thinking creatively about the world.

## Outline

**I.** Density is a ratio of the mass of an object divided by its volume. The result is a measurement, such as one gram per cubic centimeter.

    **A.** We can also talk about different kinds of density, such as population density, which involves area, not volume. In determining population density, we ask, "How many people per square mile live in this area?"

    **B.** The range of mass density in the universe is quite large, going from almost nothing to 40 powers of 10.

**II.** Some common forms of matter in the universe have extremely high density.

    **A.** A white dwarf, for example, is the hot, dense core of a star that has been released into space in the dying days of the star. This core was once the center of thermonuclear fusion for the star. Its density would be the equivalent of the density of the Sun if it were compressed into the volume of Earth.

    **B.** Neutron stars, such as the Crab Nebula, have even higher density than white dwarfs. A neutron star is the remnant of a star that exploded at some time in the universe and spread into the galaxy, enriching the galaxy with heavy elements—the active ingredients of life, planets, comets, asteroids, and so on.

        **1.** In the center of the Crab Nebula is a pair of stars, one of which was the exact center of the explosion. That star is the neutron star.

**2.** The density of the neutron star would be the equivalent of the density of the Sun if it were compressed into an area about 12 miles across. A thimble-full of the material from a neutron star would balance on a scale with a herd of 50 million elephants.

**III.** What happens under conditions of extremely high density?

    **A.** If an object is small and has a lot of mass packed into its area, it has a high surface gravity, which wreaks havoc on its immediate environment.

    **B.** If a gas cloud comes too close to a highly dense object in the universe, the gravity of that object draws the cloud in. The cloud spirals around the central point, inner regions spinning faster than outer regions, creating friction and heat. The result is high temperature and high luminosity.

    **C.** This process is a system by which these small, dense objects consume matter. The object is so small that the matter doesn't fall straight down on it but hits the spiral area, or *accretion disk*, on the side. The disk serves as a way to release energy that then descends into oblivion.

**IV.** Let's examine the densities of some other materials on Earth.

    **A.** Water has a density of one gram per cubic centimeter. We know that frozen water is slightly less dense than liquid water because ice floats.

    **B.** Frozen methane, ammonia, and carbon have about the same density as water. These materials make up comets. When a comet travels too close to the Sun, the heat evaporates these materials and helps make the tail, but the core of the comet is made up of these frozen gases.

    **C.** Rocks range in density from two to five times that of water, and Earth's crust is mostly rock.

    **D.** Metals, such as iron, are two to three times the density of rock. We find, for example, some iron on Earth's surface, but most iron has traveled down to the core of the Earth.

        **1.** In the early stages of Earth, when it was still partially molten, heavier things fell to the center and lighter things stayed on top.

2. We also know that the heavier materials are in the core of Earth from earthquake measurements. When earthquakes send seismic pressure waves through the Earth, measurements can be made of the angles of refraction of these waves to construct a profile of density for the Earth.

3. The density of the crust of Earth is about three grams per cubic centimeter; the density of the core, 12 grams per cubic centimeter; and the average density, 5½ grams per cubic centimeter.

E. Other metals, such as platinum, gold, and iridium, are much denser than iron. Osmium is one of the densest metals we know of. A cubic foot weighs about as much as a Buick.

V. Let's look at some peculiarities of density, in our thinking about it and in reality.

A. We usually say "heavy" when we mean "dense," but in some cases, that phrasing fails us. In the grocery store, for example, you see skim milk, half and half, and heavy cream, but heavy cream is lighter than skim milk—cream floats on top of milk.

B. The *Queen Elizabeth II* weighs 70,000 tons, but if it were not lighter than water, it would sink. Its total mass divided by its total volume, then, is less than one. The same is true of battleships and aircraft carriers.

C. We say on Earth that hot air rises, but that only happens if the environment has gravity. If there were no gravity, hot air would stay in the same place. When air is heated, it becomes less dense, and less dense things rise and more dense things sink.

D. Dead fish are less dense than live fish. If a live fish is neutrally buoyant, it has a density of one, and a dead fish, which floats, has a density of less than one.

E. Saturn is the only planet with an average density that is less than water. A scoop of material of Saturn would float in a bathtub.

VI. The point of infinite density is in the center of a black hole.

A. The size of a black hole is described as the size of its *event horizon*, or the boundary from which an object can never return because it would have to exceed the speed of light to do so. Black holes are called "black" because even light can't get out at the speed of 186,000 miles per second.

**B.** If black holes consume material, the event horizon gets bigger. What happens to the material? The material continues to collapse—there is no known force to prevent the continued collapse—until all the matter consumed ends up at a singular point at the center of infinite density. The center of a black hole, then, is called the *singularity*.

**C.** All the laws of physics that describe a black hole lose their applicability when we carry the matter down to the center of the black hole because of the density.

**D.** We need a new theory of physics to explain the singularity. We haven't yet replaced Einstein's theory of general relativity, which gave us black holes in the first place.

**VII.** Let's conclude with a few more mysteries of density.

**A.** Imagine for a moment that you have a box of marbles. If you add more marbles to the box, you've increased the volume and the mass of marbles—both terms in the equation for density.

1. If one term increases at the same rate as the other, the density remains the same. A small box of marbles has the same density as a large box. One has more volume and weighs more, but they both have the same density.

2. Is the same principle true with other materials? Imagine a box of down feathers. Calculate the volume and the mass, and then add that same amount of feathers to the box. Is the volume twice as much?

3. No, the volume is not doubled, because the feathers at the bottom of the box feel the weight of those on top and are squashed. The act of adding feathers to feathers makes the whole denser. You can double the mass and not double the volume.

**B.** Earth's atmosphere has the same characteristic as the box of feathers—it is compressible. The lower atmosphere is under much higher pressure than the upper atmosphere. Half of all the molecules of the atmosphere are below three miles.

1.  Astronomers try to make measurements on mountaintops or in space in an effort to get above as much air as possible so that nothing interferes with their observations.
2.  As mentioned earlier, the Earth's atmosphere extends for thousands of miles, to the point where its density equals that of interplanetary space.
3.  The space shuttle flies at 200 to 400 miles above the Earth, but even there, it still plows into atmospheric molecules, which slow down its orbit. The space station must maintain extra supplies of fuel to keep itself boosted so that it doesn't fall out of orbit.
4.  If the space station were to fall a little, it would descend into a region of Earth's atmosphere that has a much higher density of particles, which would, in turn, slow its orbit a little more and make it fall faster, and so on. The decay of the orbit is exponential.
5.  In addition, every 11 years, the energy level of the Sun kicks up as it goes through the *solar maxima*. The evidence of this is sunspots, extra solar flares and prominences, and other turbulence on the Sun's surface. This activity warms our atmosphere and makes it swell up, putting satellites at risk for having their orbits decay.
6.  When the Russians wanted to get rid of the *Mir* space station, they faced the rockets that were normally used to keep the space station buoyant in the opposite direction. The rockets were then burned to slow down the orbit and drop the ship to a lower altitude, where it was exposed to more air and descended even further. This de-orbit burn was done with such precision that the Russians were able to drop the space station straight into the center of the Pacific Ocean.

VIII. Creative thinking about density and how it reveals itself in different materials can be a valuable tool in examining phenomena in the universe.

**Suggested Reading:**

Tyson, Neil deGrasse, Charles Liu, and Robert Irion. *One Universe: At Home in the Cosmos*. Washington, D.C.: Joseph Henry Press, 2000.

**Questions to Consider:**

1. What class of cosmic object makes the most perfect spheres? Why?

2. How do the densities of the Sun, humans, and the Jovian planets compare with each other?

# Lecture Three—Transcript
## On Being Dense

Welcome back to My Favorite Universe. We're following up on an entire lecture devoted to that which is rarefied in the cosmos. I'm going to talk about that which is dense, and more generally, a discussion of what it means for things to have density at all. Some of my favorite objects in the universe have some of the highest densities of any. The range of measured densities in the cosmos is mind-boggling. It spans 40 powers of 10.

It reminds me of when I was a kid. What is the first way that we understand things that are dense and things that are not? There's the old joke, "Which weighs more, a ton of feathers or a ton of lead?" You say, "Oh, of course, it's a ton of lead." Well, of course, no. They weigh the same, because they're both a ton. The fact that some people answer that question wrong is indicative of the fact that many people think of the weight of things, not so much for how much it weighs, but for how dense it is. There's some intuitive sense of density.

We can define density mathematically. It's just a ratio of the mass of the object divided by its volume. It's very simple. That is the density. It could be grams per cubic centimeter. It could be kilograms per cubic meter. These are densities—a unit of density.

There are other kinds of densities that aren't strictly measured that way. One of them is the human resistance to common sense. That's a kind of density. He's dense or she's dense. We all know what that means, although if you took it literally, it means they have more gray matter than others with the same size head. If they're denser, it means they have more brains, so maybe there's an occasion to reverse that insult. Another way to think of density is population statistics. You're not thinking of a volume density, you're thinking of an area density. How many people per square mile live here? In Manhattan, my home borough of New York City, at mid-afternoon there's 100,000 people per square mile in Manhattan. That's very high. My wife is from Alaska, which has nearly zero people per square mile. The whole state has a half a million people tops, a very low-density population. There are other ways to think about the term "density."

In the universe, again thinking about mass density, the range is so large. We hinted to some of the thinner densities in the previous lecture. Intergalactic space has the lowest density of all, one atom per cubic meter. You can't get much lower density that that. At the other extreme, one very common form of high-density matter is a white dwarf. Here we have a white dwarf, NGC 3132. In this image—what you see here in this nebulosity is the shell of a star that had been released into space in its dying days, laying bare the hot, dense core of what was once the center of thermonuclear fusion. It is densely packed atomic matter. It's small and it's hot enough to be glowing white. We call it a *white dwarf*. To get the density of a white dwarf, what you need to do is take the Sun, or slightly less mass than that, and cram it down into something the volume of the Earth. That's how high the density is of a white dwarf.

It's not the king of high-density objects. That goes to my favorite object, the neutron star. Neutron stars can be seen in the center of this image. This particular image is of the Crab Nebula. The Crab Nebula is the remnant of a star that exploded sometime ago and spread its "guts" into the galaxy, enriching the galaxy with heavy elements. The kinds of elements you and I are made of, the kinds of elements you find that are the active ingredients of life and planets and asteroids and comets. We'll learn more about that in an entire lecture devoted just to that subject.

What's of interest here is not the crab-like nebulosity of the Crab Nebula; but, in the center of the Crab Nebula, there's a pair of stars. One of those is the exact center of that explosion and that is the neutron star, an object so dense—use the Sun analogy, take the Sun, cram it down into a volume about a dozen miles across—far denser than a white dwarf. Let's get a more earthly analogy. Let's take a thimble full of that neutron star—in fact, we gave this example in the "On Being Round" lecture—take a thimble full of it and ask, "What will it take on Earth to sort of balance that on a scale?" You put it on one end of a seesaw, and on the other end of the seesaw stack a herd of 50 million elephants, then it will just about balance. That's how crammed together the material is in a neutron star. That's dense. That's as impressively dense as anything I have ever thought of.

There are things that go on when you're dense. If an object is small, dense and has a lot of mass packed in there, it has a high surface gravity. High surface gravity wreaks havoc on its immediate environment. If a gas cloud comes a little too close, it will take the gas cloud and draw it in, in such a way that it will spiral around the central point—inner regions spinning faster than the outer regions—creating friction. When you create friction, you have heat. When you have heat, you have high temperature and high luminosity. You have these spiral areas that look like toilet bowls. We have an official word for it; it's called an *accretion disk*. It's a system by which these small, dense objects consume matter, because they're so small that matter doesn't fall straight down on it. Most of the time, it misses and hits the disk on the side. The disk is a way to release energy so that it can descend into the oblivion. Not only do you find accretion disks around neutron stars, but you'll also find them around black holes—another favorite object in the cosmos that is small and has a high surface gravity.

The density of things in the cosmos, as I said earlier, ranges 40 orders of magnitude. Let's start with simple things that we know and love and understand. Let's start with water. Water has a density of one gram per cubic centimeter. It was originally defined that way. Water is a very important part of our culture. It makes sense that we would find metric measures based on it. Other things have about the density of water: frozen methane, frozen ammonia. Frozen water has slightly less density than liquid water. We know that because ice cubes float. Icebergs float, not much above water, but they float. Ice sheets in the Polar Regions are floating on top of liquid oceans. There's a moon of Jupiter, Europa, which has ice sheets that shift around. We think there's liquid water underneath from that evidence; that all floats.

There are frozen objects in outer space. It's easy to freeze methane and freeze carbon dioxide. Comets are made of this stuff. When they come too close to the Sun, the Sun evaporates them and helps to make the tail that we know and love; but the core, the actual chunk of comet in the head of the comet, is made of these frozen gases, approximately the density of water.

On Earth we have rocks. I could have had a picture of any rock—because I'm not a geologist—so one rock is the same as another rock. Here we have a rock formation in the Southwest. What kind of density do rocks have? Rocks range anywhere from twice that of water maybe up to five times that of water. Earth's crust is mostly rock.

Let's keep going up. Is there anything denser than rock? Sure there is; there are metals more dense than rock. Iron is two or three times the density of rock. I have a sample of iron right here. This slab of iron is a meteorite. It came from Canyon Diablo meteor crater in Arizona. It's a fragment of the larger meteorite that made that crater. This is very heavy. It is very dense. It's hard for me convince you of that until I do this (drops meteorite fragment with a loud bang.) This weighs 15 pounds, about three five-pound bags of sugar; imagine that. It is iron—90 percent iron, 10 percent nickel—and a 4.6 billion-year-old piece of the origin of the solar system. Iron.

There is some iron on Earth's surface, but do you know where most of the iron went? It went down to the core of the Earth, because iron is heavier. It is denser than rock. We have iron and nickel in our core, rock on our surface and air above the rock. In the early stages of Earth, when it was partially molten, the heavy things fell to the center, the lighter things rose to the top. There are still some mixtures among them, but the core is predominantly this. The crust is predominantly rock.

There is stuff denser than iron out there. Platinum is much denser than iron, so is gold. One of my favorite scenes when you watch movies is when there is some gold heist, and they finally get into the vault; they just grab the gold bricks and they toss them to each other and walk out with their backpack full of glisten. No, no, no. As heavy as this meteorite is, gold is heavier, denser. If they're trying to steal some bricks of gold, they're going to have to take one at a time. They're not running out with satchels full of this stuff. Iridium is also very high density. The densest thing out there, however, is osmium. It would make the world's best paperweight, although this would do pretty good, too (meteorite). By the way, a cubic foot of osmium weighs about as much as a Buick, if you want to know how dense these things are.

Another way we know—you can reason that Earth would have its heavier materials in its center—but there are other ways we know this, because of earthquakes that take place periodically on Earth's surface. It sends these seismic pressure waves through the Earth. Depending on the density of material, it will refract the seismic wave at one angle or another, depending on where the earthquake was and what the run of density is as you go from the crust down through the center. If you have enough of these stations measuring the time delays, you can construct what the profile of density is for the Earth. What do we get? The core has a density of about 12 times that of water. It's up there with the heavy metals. The crust has a density of about three. The whole average for Earth is about five and a half grams per cubic centimeter, about five and a half times that of water.

As we've already noted, we so often use the word "heavy" when we mean "dense." There are plenty of things you would pick up in your life that weigh much more than this. What's a good example? A nice, well-stocked bag of groceries is going to weigh more than this. You're not going to say, "Oh my gosh!" You don't freak out by lifting it, because the size seems commensurate with the weight. That's not so with this. This is some of the most amount of dense material anyone would ever pick up. Usually when we have iron in the world—extracted from the iron ore in our crust—we would extrude this into I-beams and things where you would get strength but with low mass. We hardly ever build anything with slabs of iron this large, or slabs of steel this chunky.

Getting back to the old feathers-and-lead question, if I just come out and tell you lead weighs more than feathers—even though that sentence is not scientifically precise—you'd know what I mean by that. You know I mean lead weighs more than feathers. You're not going to say, "You didn't say that right. I don't know what you mean. You're confusing me." You know what I mean by that. I mean lead is denser than feathers. That's really what I mean. Fine. There's no ambiguity there; but, if you want to play that game, watch out. There are cases where that phrasing fails you.

I'll give you a good example. If you go to the dairy section in the store, there's skim milk, there's half 'n' half, and what else is there? There's heavy cream. The word "heavy" is labeled on the carton. Have you ever done the experiment? Heavy cream is lighter than skim milk. It floats on skim milk. Those of you who are older out there in the audience will remember that they would deliver milk, unhomogenized, to your house and the cream was on the top. The only way heavy cream can go to the top is if it is lighter than the rest of the milk that's below. When you skim it off, what is left is, of course, skim milk—because you stole all the cream for your strawberries. It fails there.

Here's another example. The *QE II*, one of the biggest ocean liners the world has known, weighs 70,000 tons. It's lighter than water. Of course it's lighter than water; otherwise it would sink. It doesn't sink. The density—total mass, divided by the total volume—is less than one. The same is true of battleships and aircraft carriers and everything else that is out there that is really massive and heavy.

I have fun with density. Let me give you some other density tidbits. Are you ready? We say that on Earth, hot air rises; but that would only happen if there was gravity. Hot air would not know where to go where there is no gravity. Out in space in the space shuttle, when they are weightless and orbiting around the Earth, if air is hot it just stays there unless they blow on it. Here on Earth we say hot air rises, but really, you could just as legitimately say cold air sinks. It's not simply that it's hot; it's that, when you heat air, it becomes less dense and less dense things rise and denser things sink. I'd like to sort of broaden the vocabulary and spread the word. Instead of saying hot air rises, cold air sinks.

What else do we have? I've got a good one. We already know that solid water is less dense than liquid water. Here's another one: dead fish are less dense than live fish. How do we know this? Because they float, belly up, for at least some part of the time after they have died. The other fish are just happily moving around. If the fish is sort of neutrally buoyant, it means a live fish has a density of one and a dead fish has a density less than one; hence, the posturing with its belly up in the tank.

We mentioned this in an earlier lecture, but it's worth repeating, that the density of water, the density of human beings, and the density of the Sun are all about the same. The Sun is a little bit denser, but it's all about the same: one gram per cubic centimeter. I'd like to think so many people flock to the beach during the summertime and we so enjoy water, that there is some commuting going on. I think we know that we're the density of water, and we're made mostly of water and we want to sort of get close to it. There's something deeper going on than just, "Oh, I want to play in the sand."

My favorite planet, the planet Saturn, has a distinction among all planets in the solar system for being the only one whose average density is less than water. If you had a bathtub, and you took a scoop of the average material of Saturn, it would float in your bathtub and it's the only planet that would do that. When I was a kid, instead of the rubber ducky you have to play with in the bath, I wanted a little rubber Saturn, because I knew that Saturn would float. Alas, no one has come up with such a product, but I encourage you to, out there in audience land.

There is a limit to density, even beyond that of neutron stars. Here's what happens at the center of a black hole. The size of a black hole, by convention, is described as the size of its *event horizon*, which is that boundary beyond which an object can never return, because it would have to exceed the speed of light to do so. That's why they're called "black," because light can't even get out—even at the breakneck speed of 186,000 miles per second. So it's black. If you cross the event horizon, you're never coming out.

If black holes "eat," then the event horizon gets bigger, bigger and bigger. What happens to the material? The material keeps collapsing down, and there's no known force to prevent the continued collapse until all of that matter ends up at a singular point at the center of infinite density. We call that, sensibly, the *singularity*, the center of a black hole.

I don't even know what "infinite density" means, but I can tell you that our laws of physics that describe a black hole—when we carry the matter down to its center—all those laws of physics lose their applicability. We're in desperate need of a new theory of physics in order to explain the singularity. When I say it has infinite density, I'm just professing my ignorance, because the community has not yet come to a replacement theory for the general relativity of Albert Einstein that gave us black holes in the first place. There's a missing piece. Stay tuned for that. That's infinite density.

There are some mysteries to density. My favorite mystery is that a can of Diet Pepsi floats, and a can of regular Pepsi sinks. We may never understand that one. By the way, the only way you would know this—because there aren't that many occasions to float cans of soft drinks—is, if you hang out very late at a party when everyone else has gone home and the big cooler that used to be filled with ice is now water, with the remaining ice cubes afloat on top. By whatever cans of soft drinks are left, you get to determine whether they have sunk or whether they are floating. Take a look. You'll see the Diet Pepsis floating and the regular Pepsis have sunk. That is one of the most profound mysteries I know.

Let's look at ways that things can behave. The black hole is odd because it eats—all the material goes to a singular point of infinite density—and the event horizon grows. That's kind of peculiar. Let's think about the classical way you would add material to something. Let's think of marbles in a box. If you add more marbles, you've increased the volume and you've increased the mass. You've increased both terms in the equation for density: mass divided by volume.

If one increases at the same rate as the other one, the density is the same. It just divides out. A small bag of marbles has the same density as a large bag of marbles. One weighs more. It has more volume, but it's got the same density. Fine.

Would that happen with other ingredients? Those are marbles, but suppose we took down feathers. Let's try that experiment. Dump a bunch of down feathers in a box, figure out how much mass it is, and look at its volume. Now take that same amount and add it to what you just put in that box. What is the volume? Is it twice as much? No, of course not, because it's down feathers and the down feathers at the bottom are feeling the weight of the extra feathers on top. They're getting squished. The act of adding down feathers to a previous supply of down feathers makes the ensemble denser than how it started. You could, in fact, double the mass and not double the volume. That's a general behavior of squishy things.

Earth's atmosphere is squishy; it's compressible. The lower atmosphere is under much higher pressure than the upper atmosphere. Half of all the molecules of Earth's atmosphere are below three miles, half of all the air. That's why astronomers are always running to mountaintops trying to get above as much air as possible. The next best thing to do is to launch something into orbit where you are above 99.99999 percent of all the air molecules that could interfere with your observations.

Where does Earth's atmosphere end? There's not a signpost that says, "You are now leaving Earth's atmosphere." NASA suggests that they leave the atmosphere. They suggest so because they say, "Oh, we're up in orbit and you see clouds way below you." Of course, it makes sense to think of it as such, but no. No, Earth's atmosphere goes for thousands of miles, and only at thousands of miles does the density equal the density of interplanetary space. Then you could say, "I can't tell the difference any more between being closer to Earth and being out there between the planets." That must be the edge of Earth's atmosphere, thousands of miles out, even at the distance that the space shuttle flies and the space station—they fly, anywhere between 200 and 400 miles up. Surely there's no atmosphere there.

That's not the case. Here's the space station with all its solar panels in orbit around the Earth; and up there, even though it's a couple of hundred miles up, there still remain atmospheric molecules. This thing is plowing into them, and what effect does that have? It slows down its orbit. In fact, the space station and every object in lower Earth orbit, a couple hundred miles up, has to go up with extra fuel to keep itself boosted so that it doesn't just fall out of orbit; because, once it starts falling, there's no stopping it. If it were to fall a little bit, it would descend to a region of the atmosphere that is a much higher density of particles than there once were, so it would fall faster. If it falls a little bit more, then it's in a denser area and it would fall faster yet. The decay of the orbit is exponential.

Do you know when it's especially worse? You may have heard or remember that the Sun goes through cycles, 11-year cycles of activity. Every 11 years the energy level of the Sun kicks up a notch, and then it descends back down. Evidence of that are sunspots, extra solar flares and prominences and all kinds of turbulent features on the Sun's surface. That extra activity warms Earth's atmosphere, and the act of warming it makes it swell up and reach out farther from Earth's surface than it normally does.

Anytime you're going through *solar maxima*—as we call it, *solar max*—all of the satellites orbiting between 200 and 400 miles are at higher risk than they would otherwise be for having their orbit decay. You have to be careful when you launch satellites. You have to watch for the density of the atmosphere and watch for what the Sun is doing. It's not insurmountable. You just have to give a periodic boost, but you have to know that you have to do that.

One of my favorite cases of this is the largest thing ever de-orbited. You may remember. It was a few years back: the *Mir* space station. The *Mir* space station was built by the Soviet Union followed on by the Russians after 1989. The *Mir* space station was getting long of tooth. The Russians agreed to participate in the international space station, which is up there now, the first permanently occupied space platform. What did they do with *Mir*? They had to de-orbit it. How did they de-orbit it? The rockets that were normally used to keep the space station buoyant were faced in the opposite direction. The rockets slowed down the orbit, dropped it to a lower altitude where it was exposed to more air, and it descended rapidly.

It needed to be done with such precision; plus, fortunately, there is a big, wide, open area called the Pacific Ocean. The de-orbit burn was started out over Asia, and by the time that thing came tumbling out of orbit, it was aimed straight for the center of the Pacific Ocean so it wouldn't fall on somebody's head. A lot of pieces burned up in the atmosphere; the rest just descended into the bottom of the Pacific. We think of it as littering the Pacific, but there's no comparison to how many lost ships from the past 400 years are at the bottom of the Pacific Ocean. Our little space debris is nothing compared with how many lost voyages of explorers of the past 400 years we can find at the bottom of all the oceans, even in areas very close, like Cape Cod Bay, for example.

The lesson here is you need to think creatively about the concept of density and how it plays out in your life. It's not just some abstract scientific concept. There are very valuable tools you can bring with you by thinking of density, thinking of how density reveals itself in different materials. Take, for example, an automobile accident. Some cars survive accidents better than others. In almost every case, it's because they not only weigh more but they're also denser. The denser the material, the more likely it is to survive an encounter with something else that is passing through. Little things like that.

I'll leave you with this one case. I was in a restaurant in Pasadena, California. It was one of these coffee houses that also served evening dessert. You go to your fancy restaurant then you do your fancy dessert at one of these places. One of my favorites is hot chocolate with as much whipped cream as could possibly be balanced on top. I ordered the hot chocolate with whipped cream, and in came the hot chocolate and I didn't see any whipped cream at all. I asked the waiter, I said, "Waiter, where's the whipped cream?" Do you know what the waiter said? I couldn't believe it. The waiter said, "Oh it must have sunk to the bottom." I'm thinking, "Either the laws of physics are different in this restaurant from the rest of the universe, or you are mistaken, Mr. Waiter." He said, "Well I'm sure. I'll show you." He went back and got blobs of whipped cream, came out, and plopped it in. You know what it did? It sunk for a split second, because it had momentum, then popped right back up and sat up. I said, "Thank you for putting whipped cream on my hot chocolate."

# Lecture Four
# Death by Black Hole

**Scope:**

Black holes seem to be one of the most fascinating topics in the universe. In this lecture, we'll learn what they are, how they would kill a human being, and how they wreak havoc in the universe. We'll also touch on recent research suggesting that every galaxy has a black hole at its center and how that fact affects our conception of the universe.

## Outline

**I.** A black hole is a region of space in which the escape velocity exceeds the speed of light.

    **A.** Escape velocity is the speed at which an object must be launched for it to escape its environment forever. For a black hole, the escape velocity is the speed of light, which means that even light cannot escape.

    **B.** Escape velocity is correlated with gravity. Objects that have higher gravity have higher escape velocities.

    **C.** On Earth, the escape velocity is seven miles per second, or about 25,000 miles per hour. A low-mass object, such as a comet, has an escape velocity of one meter per second, or two miles per hour.

    **D.** The Moon has an escape velocity of about 2.5 kilometers per second, or about 1 to 1.5 miles per second. Astronauts had to be launched from the surface of the Moon at that speed; otherwise, they would have fallen back. The Sun has an escape velocity of about 600 kilometers per second, or 400 miles per second.

    **E.** The mass of the object for which we are trying to determine the escape velocity matters, because objects with high mass have high gravity. Size matters, as well. The more compressed an object is, the closer the surface of that object is to its own center. The force of gravity is related both to mass and distance to the center.

**F.** A black hole might not even have much mass—maybe just a few times the mass of the Sun—but because it is small, its surface gravity is high and its escape velocity is high.

**G.** The point of no return from a black hole is called its *event horizon*.

**II.** We discovered black holes through Einstein's general theory of relativity.

  **A.** This theory describes motion, including gravity and the acceleration of gravity.

   **1.** In this theory, gravity is not just a force of attraction between two objects. Gravity curves space. When objects move, they are moving in response to the curved fabric of the universe.

   **2.** Imagine that we live in a two-dimensional universe that is something like a sheet of rubber. Just as that sheet could be warped, the gravity in our three-dimensional universe can be warped.

   **3.** The warp forms a funnel shape in the fabric, and its center would have very high gravity.

   **4.** Objects in orbit cause even more distortions in gravity.

  **B.** A student of Einstein, John Archibald Wheeler, summed up the general theory of relativity by saying, "Matter tells space how to curve; space tells matter how to move."

**III.** Why would matter collapse to form a black hole in the first place?

  **A.** Usually, matter is supported against collapse from gravity. For example, when an object is heated, its atoms or molecules are in constant motion, creating a pressure that resists gravity and prevents collapse. If the object has a high enough mass, however, its gravity can overcome the sustaining gas pressure.

  **B.** Matter can be squeezed together until it is in a state in which atoms are right next to each other. This is what forms a *white dwarf*. Matter can be squeezed even closer until nuclei are right next to each other, forming a *neutron star*, which is the densest matter we know.

  **C.** What happens if the force of gravity is so high that even the pressure of neutrons can't support against it? We know of no

©2003 The Teaching Company Limited Partnership

force of nature that can support an object against such gravitational forces. Once that object exhausts its supply of nuclear fuel and begins to collapse, nothing is available to support it against collapse.

**D.** The matter in this object descends through its own event horizon. As far as we know, all the mass collapses down to a single point of infinite density and zero volume. That point is known as the *singularity*. General relativity—the theory of gravity—fails at the singularity.

**IV.** Let's dive feet first into a black hole.

**A.** If you're falling through space and you begin to descend toward a black hole, the force of gravity grows exponentially. You would start to fall faster, but if you're in a free fall, you're weightless, which means that you would not even notice the acceleration of your fall.

**B.** What kills you in your descent toward the black hole is not gravity, but the difference in gravity felt between your feet, which are closer to the black hole, and your head.

   **1.** On Earth, the force of gravity is also greater at your feet than at your head, but you don't feel this difference, because your size is tiny compared to that of Earth.

   **2.** Black holes, however, are themselves tiny, and this size magnifies the difference between the force of gravity at your feet and at your head.

**C.** As you descend toward the black hole, you would begin to stretch under *tidal forces*. The same force that pulls at the oceans on Earth from the Moon would also stretch you.

**D.** When the tidal forces exceed the chemical bonds of human tissue, you would be torn in half. Then, those two pieces of your body would be torn in half, and so on. Eventually, you are completely torn into countless pieces of biological matter as you descend.

**E.** Further, in your descent, you are occupying a space that is getting narrower and narrower, again, like a funnel. Thus, not only are you stretched head to toe, but also you are also squeezed shoulder to shoulder.

**F.** The result is similar to what happens when you feed dough into a pasta machine. In fact, what happens to matter as it descends into a black hole is called *spaghettification*.

**G.** Keep in mind that black holes get bigger in exact proportion to how much mass they consume. They can be any size, which is defined as the size of the event horizon.

    **1.** Only small black holes will kill you before you descend through the event horizon.

    **2.** For large black holes, the tidal forces at the event horizon are relatively low and, therefore, less damaging to human tissue.

    **3.** A low-mass black hole—the smallest—causes the worst damage, because the rate of change of gravity becomes significant as you near the event horizon.

    **4.** In either case, you'll be torn apart. If you're falling toward a small black hole, it will happen before the event horizon; if you're falling toward a large black hole, it will happen after you reach the event horizon.

**V.** How do black holes wreak havoc in the universe?

**A.** Stars frequently travel in pairs, called *binary stars*. As a star in one of these pairs starts to age, it swells up and becomes a red giant. As it gets bigger, some of its material might get too close to neighboring objects, such as a black hole, which will consume it.

**B.** In this phenomenon, we usually see a disk of material around the black hole, called the *accretion disk*. This is the collection area of gas from the red giant that is feeding the black hole.

**C.** We also see jets spewing out above and below the black hole. So much material from the red giant is descending into the black hole that it becomes heated up by friction. The resulting energy must escape in these jets.

**D.** At these high temperatures, the disk starts to emit ultraviolet light and x-rays.

**E.** Some black holes in the centers of galaxies are massive, as much as a billion times the mass of the Sun. Such black holes consume whole star clusters. The accretion disks are huge and emit significant radiation. In fact, such an accretion disk can outshine the entire galaxy in which it is embedded.

**VI.** Black holes were discovered because of the radiation they emit, along with radio waves, all stemming from one tiny spot.

    **A.** At first, scientists didn't know what black holes were. Because the energy profile of a black hole didn't quite match that of a star, they were called quasi-stellar radio objects— *quasar*.

    **B.** A quasar can be understood as a galaxy with a supermassive black hole. It has such high energy coming from such a small spot that it is visible as a star, but it really exists at the edge of the cosmos.

**VII.** Can a black hole shut down?

    **A.** The event horizon can stretch out so far that the black hole is no longer able to rip things apart, because the tidal forces become very shallow.

    **B.** If the black hole is that large, it consumes material whole and no radiation is emitted. It is possible for a black hole to get so large that it shuts off its mechanism.

    **C.** A black hole can also shut off if it consumes everything that is near it.

**VIII.** Some evidence suggests that all galaxies have black holes at their centers, some more massive than others.

    **A.** This discovery would mean that all galaxies are the same, but they have different properties, such as different masses of the black holes, different rotation rates, and so on. This conception helps us understand galaxies better, because we are starting from a core of similarities.

    **B.** The Milky Way has an "ordinary" black hole at its center. Our black hole is about a million times the mass of the Sun. It is not big enough to have ever looked like a quasar.

**IX.** We'll close with a literary tribute to black holes:

        In a feet-first dive

        To this cosmic abyss,

        You will not survive,

Because you surely will not miss.

The tidal forces of gravity

Will create quite a calamity

When you're stretched head to toe.

Are you sure you want to go?

Your body's atoms—and you will see them—

Will enter one by one.

The singularity will eat them,

And, of course, you won't be having fun.

**Suggested Reading:**

Shipman, Harry L. *Black Holes, Quasars, and the Universe*. Boston: Houghton Mifflin Co., 1976.

Thorne, Kip. *Black Holes and Time Warps*. New York: W.W. Norton & Co., 1994.

**Questions to Consider:**

1. What special feature of small black holes makes them more dangerous than large ones?

2. When an object collapses and occupies smaller and smaller volumes, what happens to its surface gravity? Does it go up, down, or stay the same? Why?

# Lecture Four—Transcript
## Death by Black Hole

Welcome back to My Favorite Universe. We're going to talk about black holes. This subject is one of the most fascinating in all of the cosmos. By way of statistics, when I'm out in the public and people know that I'm an astrophysicist, it's one of the top three questions I get asked. Usually I get asked something about the search for life in the universe, next is about the Big Bang and third, right up there with those two, is what is a black hole? Are they dangerous? What will it do to me if I run into one?

First of all, yes, they are dangerous. They will kill you posthaste, and they're also extremely fascinating because they wreak havoc on the environment in which we find them. I've titled this lecture "Death by Black Hole." We can first describe how a black hole would kill a person—that's fascinating unto itself—but, this havoc that gets wrought upon the environment by black holes throughout the galaxy and throughout the universe, in a way, is killing stars and gas clouds. So, it's really just a lecture about death and destruction caused by black holes. For human beings, it's quite a spectacular way to die. If I were to choose a way to go, I'd say launch me into a black hole. What it does is it rips you apart, atom-by-atom, and you go in as a stream of matter right down into this bottomless abyss. We'll detail that a little more in just a moment.

What is a black hole? Let's start at the beginning. A black hole is a region of space within which the escape velocity has exceeded the speed of light. The escape velocity is the magic speed at which an object must be launched for it to escape its environment forever. It makes sense that higher gravity objects have very high escape velocities, but for a black hole, that escape velocity is the speed of light. The speed of light is the fastest stuff we know of in the whole universe. If light can't get out, nothing gets out; hence the name, "black hole." It's black because nothing comes out; it's a hole because if you fall in, you're gone. It's sensibly identified.

Let me give you a sort of grassroots discussion of escape velocity. I'll use my shoe for this demo, if you don't mind. If I take my shoe and play catch with myself—if I toss it up, it goes up about a foot and comes back. I threw it at a particular speed in order to reach one foot above my head. But, if I threw it at a higher speed, it goes even higher. It takes longer for it to return to me. If I keep doing this, constantly increasing its speed, I can do the experiment and find out at what speed will the shoe never come back. If you do that experiment, you'll find that that speed is about seven miles per second. Converted into speed-limit language, it is 25,000 miles per hour—extremely fast. You don't encounter this in everyday life. That's probably what led to the adage, "What goes up must come down." But that adage is more just for some everyday experience. If you look at the physics of the cosmos, there is a speed at which you can launch something so that it will never come back, rendering that childhood-remembered phrase incorrect. You might go back and tell your teachers that.

It makes sense that this escape velocity must somehow be correlated with how much gravity there is on the orb from which you are launching your object. Let's take a very low mass thing, like a comet. Comets are small. A big comet might be 100 miles across. The escape velocity on a comet is about one meter per second. You can calculate that out. That's about two miles an hour. You can walk that fast. A brisk walk is three miles an hour. If you ever found yourself on a comet and just started walking fast, watch out, because you will propel yourself into orbit around the comet. By the way, that's why comets are very dirty things in the solar system. If anything gets jostled loose on it, it flies away and the comet loses it. It's this big garbage heap and it strews garbage behind it every time it comes around the Sun. That's the cause of meteor showers. Earth plows through the debris of these things and it rains down on Earth's surface. It's because it has a low escape velocity.

What's next? We can go to the Moon. The Moon has more mass and more gravity than a comet. For the Moon, the escape velocity is about two and a half kilometers (2.5 km/s) per second. That's relatively fast. That's about one to one and a half miles per second. I can't throw anything that fast. You need launch vehicles to do that, even on the Moon. NASA did that, of course. We landed on the Moon then escaped back out from the Moon. They had to be launched from the Moon at that speed; otherwise, they would have fallen back to the Moon.

Let's keep going. As we know, for Earth, it's 25,000 miles per hour. The Sun escape velocity is a little more than 600 kilometers per second, about 400 miles per second. That's fast. That's really, really fast. It's very hard for things to just sort of escape the solar system, because the Sun has this near-eternal lock upon them—because of such a high escape velocity. What the number means is, if you're standing on the surface of the sun, and you want to say, "Goodbye solar system," if you don't have that speed, you're going to fall right back to the Sun from whence you came.

Mass matters. High-mass objects generally have high gravity. But, size matters, too. The more compressed an object is, the closer you can get to its center—the closer the surface of that object is to its center. The force of gravity is related not only to the mass, but also to the distance to the center. That is the force of gravity. Black holes and other dense objects in the cosmos might not even have that much mass. A black hole might have a mass a few times the mass of the Sun. There are plenty of stars with that mass, but it's because the black hole is small that its surface gravity is high, and the escape velocity is high. It can plunge down in and close back on itself, preventing light from escaping, because the escape velocity has exceeded the speed of light. We think of this region of space as a surface, but it's not a solid thing. It just happens to be within which you don't come out; it's the point of no return. We have a word for that, this point of no return, rather poetically described as the *event horizon*, the event horizon of a black hole.

You might ask, "Where did all this come from?" Did we just make this stuff up? No, no. Someone you've heard of before, Albert Einstein, introduced the theory of relativity back at the turn of the century, in 1905. That was the special theory of relativity concerning motion and straight lines, non-accelerating motion. Ten years later, 1915—published in 1916—he came out with the general theory of relativity, which described motion of any kind, accelerated motion including gravity and the acceleration of gravity. That is known as Einstein's general theory of relativity. In that, he described the force of gravity not purely as a force of attraction between two objects, that's very Newtonian—Newton's theory of gravity thinks of attractions in that way. What Einstein said was, gravity curves space; and when objects move, they're moving in response to that curved fabric of the universe.

There is something called an "embedding diagram," which indicates this for a black hole. We live in three dimensions plus a dimension of time; that gives us four, and it's hard to imagine that warping. We'd be able to see it if we were higher dimensional creatures looking down on the space in which we live. Let's do that, but for our case, let's create a two-dimensional universe, imagining a sheet of rubber with a grid on it. Let us warp that sheet of rubber. We're living inside that sheet of rubber and there's the warp that we're talking about.

That is the kind of warp that exists, except in higher dimensions, in the universe we live as described by Einstein. Here, because of our feeble human minds, we can only imagine it in the two-dimensional case. The warp forms a funnel and its center is an object of very high gravity. If it's a black hole, that funnel is so long you're never climbing out of that to escape back into the grid of the rest of the cosmos. That's a way to think about what's going on. If there were two objects in orbit around each other, you'd have two dimples. Imagine a cluster of stars; there would be dimples all over, all responding to their own distortions of space and of time.

There is a phrase I first heard from John Archibald Wheeler, who was a student of Albert Einstein, summarizing all general relativity. He says, "Matter tells space how to curve; space tells matter how to move." For me, that summarized all of the philosophical underpinnings of the theory of general relativity.

Why would something collapse down into a black hole in the first place? Usually things are supported from collapsing under gravity. There could be chemical forces, there could be thermal forces, the movement of molecules; gravity is trying to squeeze it down but they're moving fast so that prevents it from collapsing. Those are forces. But, if you have so much mass, you can actually overcome that and cram matter down until it is sort of atom-next-to-atom. That's possible. You can do that. We have states of matter that are just that. We call them *white dwarfs*. You can cram it down even more until it's not just cheek-to-cheek, atom-to-atom; but it's nucleus-to-nucleus. If you do that, you get a *neutron star*. That's the densest matter we know. Neutron stars, if they're rapidly rotating— and we see them that way—we would label them as *pulsars*.

What happens if it has so much gravity that this pressure of neutrons can't even support against it? That's all she wrote. We know of no force of nature that can support an object against the gravitational forces of, let's say a ball of mass five, six, ten times the mass of the Sun. Once that starts making its nuclear fuel and it begins to collapse, nothing is left available to support it against the collapse. The matter just keeps going and going. It descends through its own event horizon. That's when it disappears from view and, as far as we know, all the mass collapses down to a single point of infinite density and zero volume. That's kind of absurd. What does that even mean, "infinite density?" I don't know what that means, but what I do know is that general relativity, the most successful theory of gravity ever put forth, fails at that singularity. We know that it's a theory that is incomplete. Even though everything it has predicted has been true, it's an incomplete theory and we're searching. "String theory" is one of the ideas put forth that will give us a handle on the singularity deep inside.

It's for all these reasons that black holes are some of the most romanticized of cosmic objects. They're mysterious, they're dangerous and they're basically fertile for storytelling, particularly in the realm of science fiction.

Why don't we take a feet-first dive into a black hole and see what happens. As I already mentioned, as you're descending toward the black hole, the force of gravity is growing exponentially. All this means is you fall faster and, if you're in free fall, you're actually weightless. You wouldn't care that you're descending toward a black hole because you wouldn't notice it just yet. You'll just be falling. It would be like falling toward Earth or falling toward anything. That's not what kills you. It's not the high gravity that kills you.

Do you know what it is? It is the difference in gravity between your feet, which are closer to the black hole, and your head. As I stand here on Earth, there's a difference in gravity between my feet and my head. I barely notice it because my height is very small compared with the radius of Earth. You can ask, "How much stronger is the gravity at my feet than at my head." Look at the size of Earth then look at my height. That's basically nothing. I don't even notice that difference.

If you're descending toward a black hole—and black holes can be tiny things—if the size of the black hole and your height are comparable. Imagine in a limiting case, I've got a six-foot diameter black hole and I'm about six feet, and I'm falling toward the black hole. My feet are twice as close to the black hole as my head. If you calculate what force that is, my feet are feeling an acceleration toward the black hole that is four times the acceleration of my head. I begin to stretch. My body begins to feel like I'm on a rack. If I were made of rubber—if I were Rubberman—I would just stretch, according to these forces. But, I'm not made of rubber. I'm made of human flesh, and that has force limits.

By the way, we have a word for this force that stretches you as you descend toward the black hole. It's called *tidal forces*. It's the same word for the tides on Earth. Earth feels tides from the tidal force of the Moon. One side of the Earth is closer to the Moon than the other in a measurable way, so the oceans feel a little bulge. They're drawn a little more on the near side of the Earth than on the far side of the Earth, so there's this stretching of the oceans in response to the tidal forces of the Moon.

If the Moon can do that to Earth, imagine what a black hole can do. What happens? As I begin to descend, eventually these tidal forces exceed the chemical bonds of human tissue. As I descend there is a point where I can no longer resist it and I snap into two pieces, a lower segment and an upper segment. I keep falling and, sure enough, those two segments feel enough of a tidal force that they snap into two pieces. Then those four pieces each snap into two pieces, and it goes from one to two to four to eight to 16, and this just continues. Eventually, you are completely snapped into countless pieces of biological matter as you descend toward the black hole.

It gets worse than that. It turns out; space and time have collapsed onto a black hole. This is what general relativity describes for us. It squeezes down into the black hole. So here I am occupying a space that is getting narrower and narrower and narrower like a funnel. Not only am I stretched head to toe; I am squeezed shoulder to shoulder. It's as though my body is being extruded—I think of toothpaste being extruded through the hole in a toothpaste tube. That's what's going on as you fall toward the black hole.

I don't know if you've ever had one of these homemade spaghetti-making machines. You take the semolina dough; you knead it and put it up—put it in the machine. You squeeze it and out the other side come these long strands of spaghetti. In fact, this phenomenon is officially known as *spaghettification*; its what happens to matter that's descending into a black hole.

Black holes eat. What happens to them? They get bigger. It's true with most things; they eat and they get bigger. Black holes get bigger in exact proportion to how much mass they have consumed. If a black hole is of a given size and a given mass, and it eats the equivalent of its own mass again, then it becomes twice that size. If it eats three times as much, it becomes three times that size. The arithmetic of relativity demonstrates why this is so. Again the size is referring to the size of the event horizon.

What that means is that black holes can actually be any size, depending how hungry they've been in their lives. Not all black holes will kill you before you descend through the event horizon. Small ones will. But for really, really big black holes, the tidal forces right at the event horizon are relatively low. They're less damaging to human biology as you descend. What that means is it's the low-mass black holes, the ones that are the smallest, that will do the worst damage to you as an unsuspecting visitor. It's because the rate in change of the force of gravity— that's the tidal force—gets significant as you near its center.

In both cases you'll get ripped apart no matter what, because there's some distance from the center where this will happen to you. The difference is that with a small black hole it will happen before you get to the event horizon and everyone will get to see this happen. Whereas, in a big black hole, you'll descend through the event horizon, you'll get ripped to shreds and no one will know, except for you—but you won't be able to tell anyone about it, of course.

Suppose you have really, really, really big black holes. Let's go to the real universe. I was describing ones that might eat a person, but let's get to ones that do some real damage in the universe. In the real universe you have stars that don't necessarily travel alone. There are a lot of stars that travel in pairs; they're called *binary stars*. When you have stars in pairs, typically one will age before the other, and part of its aging means it becomes a red giant and swells up. As it gets bigger, some of its material gets a little too close to the neighboring object. If the neighboring object is a black hole, the black hole is going to eat it. It's not just going to eat it; it's going to flay the red giant as it expands into its space. We have an image of this. First, there's a person descending down into the black hole. That person is not having fun. You see the spaghettification in progress. By the way, that would happen to any material going in, it's just more explicitly conveyed when you're a human being. Notice in this image, there is a star, a red supergiant that has become big and bulbous. In that effort some of its material has gotten a little too close to its companion. That little companion—that little dot off to the left—has a disk of material around it that is the collection area of the red giant gas, and it feeds the material. It feeds the hole in the center. It feeds the event horizon.

If we take another look at this, go close up; we notice that there are jets coming out above and below. There is so much material trying to descend into that black hole, and it gets so hot because there is friction from all of the gas rubbing against itself—descending down into the center—that that friction heats up the gas and that energy is trying to escape somehow. It can't escape through the plane, so it pops out the top and the bottom. This is a classic image of a black hole with a disk of material around it, in companion with a red giant, and jets being spewed above the plane and below the plane.

At those temperatures that it reaches, it starts radiating ultraviolet light and especially x-rays. As you look out in the universe at a star that you think is minding its own business, it's a sure giveaway of this kind of system if you whip out an x-ray telescope and see that it's one of the brightest objects in the sky. You'd say, "Something's going on there." It's a disk of material heated, coming down to a small central area. You see the black hole in the center, and you just want to avoid that. You don't want to come anywhere near this thing.

There are black holes bigger than this. These black holes are maybe ten times the mass of the Sun, and they're wreaking havoc on its neighbor star. They get bigger than this, though. There are black holes in the centers of galaxies. We know this because we've looked at galaxies, galaxies of stars. We're talking about systems that have 100 billion stars and in their center is a super massive black hole, in some cases a billion times the mass of the Sun—a billion times. They're enormous.

These black holes can wreak such havoc on their environment that whole star clusters can get eaten. In the previous case of the blue supergiant—or it could be a red giant, any large, swelled-up star—in the case of the large star and the neighboring black hole, it's only eating one star at a time. If you go down to the center of the galaxy, this black hole is so large it has the capacity to dine on enormous gas clouds and star clusters; a voracious appetite. It, too, will make an accretion disk of titanic proportions. In fact, the radiation coming out of these accretion disks is so significant that, on occasion, it can outshine the entire galaxy in which it is embedded.

When such objects were discovered, we didn't even know what they were. We said, "What is this?" They emitted not only high-energy radiation; they also emitted radio waves. They were discovered using radio telescopes, and most of the energy came from a tiny little spot, not spread out like you'd see in a full galaxy. When these were discovered they weren't really stars because their energy profile didn't match that of stars. We called them "quasi-stellar," and they gave off radio waves, so we called them "quasi-stellar radio objects"—*quasars*. It turns out; these were galaxies with super massive black holes sitting at the edge of the universe. There is such high energy coming from them and from such a small spot, that if you look at it in the night sky you'd say, "Oh that's another star sitting up there, when in fact, the thing is hailing from the edge of the cosmos. Extraordinary.

It is possible for a black hole to get so big and to have the event horizon stretch out so far that it is no longer able to rip things apart, because the tidal forces become very shallow. When things are ripped apart, that's what makes the accretion disk, this disk of material that feeds the center. That's where the friction takes place. If the black hole is so large that objects don't get ripped apart as they descend, then they get eaten whole. If they're eaten whole, no radiation comes out, no high energy x-rays, gamma, none of that; so it's possible for a black hole to get so big that it shuts off this mechanism. We think some quasars that have turned off since the beginning of time, have done just that.

There is another way to turn off. If a black hole has consumed everything in its environment and nothing comes close to it any more. That's another way to turn off. It just runs out of food.

The latest evidence suggests that perhaps all galaxies have black holes in their centers, some more massive than others. Some galaxies that have super massive black holes, which are kind of quiet, we suspect would have looked like quasars to us in the distant past. It's a shift in paradigm of how we think of galaxies in modern times. When I first started school it was, well there is this kind of galaxy and that kind of galaxy, there's a quasar and so on. It's all the same kind of galaxy with just slightly different properties within it. It's not a different species. They're all the same species just with a different mass of the black hole in the center and slightly different rotation rates. By bringing them all together in one intellectual construct it enables you to more thoroughly examine how it is that these objects differ, because you're already grabbing onto a core of what is the same. The fact is they're all galaxies.

The Milky Way, our galaxy, has a black hole, too. No, it's not as big as the biggest. It's like an "ordinary" black hole. If you have black hole envy, our black hole is about a million times the mass of the Sun. That's not big enough to have ever really looked like a quasar in the early times. It is big enough to kind of disturb the middle of the galaxy, and there are colleagues of mine whose entire research program focuses on exactly what's going on in the center of the galaxy. We track the motion of stars that get almost too close to the black hole. The gravity is so high they get pulled in very fast. You see stars minding their own business, then whoosh. They come a little too close to the black hole and you see their speed increase dramatically; speeds that you don't find anywhere else in the galaxy. That tells you there's a lot of mass in a small volume. These are some of the ways we have deduced that every galaxy we have ever looked at has evidence for a black hole in its center.

Black holes are things to respect in the cosmos. I'd like to summarize their danger with just a rhyme. Suffer through this rhyme that I once composed, because I felt compelled to do this, because I couldn't get black holes out of my head. It goes something like this:

In a feet-first dive

To this cosmic abyss,

You will not survive,

Because you surely will not miss.

The tidal forces of gravity

Will create quite a calamity

When you're stretched head to toe.

Are you sure you want to go?

Your body's atoms—you will see them—

Will enter one by one.

The singularity will eat them,

And, of course, you won't be having fun.

# Lecture Five
# Ends of the World

**Scope:**

This lecture discusses phenomena that could bring an end to planet Earth. When we contemplate such an occurrence, we usually think of a rampant virus that would decimate our species, or global thermonuclear war, or the destruction of the environment. These tragedies, however, would result only in an end to human beings; Earth itself would still exist. We can point to three scenarios that would result in the destruction of the planet: the death of the Sun, the collision of the Milky Way and the Andromeda galaxies, or the heat death of the cosmos. We won't live long enough to see any of these, but perhaps our species will.

## Outline

**I.** The first scenario we'll examine for the end of the world is the death of the Sun.

    **A.** The Sun was born about five billion years ago, giving birth to the solar system. According to predictions of its stellar evolution, the Sun will live for about another five billion years.

    **B.** The Sun is mostly made of hydrogen and helium. Its surface is a turbulent plasma that experiences strong magnetic fields. The surface temperature is 6000 degrees Kelvin (K); the core temperature is about 15 million degrees K.

    **C.** The core, where thermonuclear fusion takes place, is stable and is the source of all energy of the Sun. There, the nuclei of hydrogen atoms are brought together to form helium atoms. This act of *fusion* results in a loss of mass and a release of energy.

        **1.** The Sun would tend to collapse under its force of gravity, but it can't because this energy is working its way out. The thermal energy released by this thermonuclear fusion, and the gravitational pressure trying to collapse the Sun, are in balance.

        **2.** Therefore, the Sun keeps its same basic shape and same rate of energy output.

**D.** The conversion of hydrogen into helium can continue only as long as hydrogen exists in the core of the Sun at certain temperatures.

1. Extremely high temperatures are needed to bring about the fusion of the two protons in a hydrogen nucleus. Because the protons are both positively charged, they would naturally repel each other. High temperatures increase the speed of the particles to the point where they overcome their repulsion and bind together. At the moment the protons touch, the *strong force of nature* takes over to hold them together.

2. When the Sun runs out of hydrogen, a mass of helium will be left in the core. Helium, however, will not fuse at 15 million degrees; even higher temperatures are required.

3. The Sun has a mechanism to increase its temperature, called *loss of equilibrium*. When no hydrogen is left to support the Sun, it will begin to collapse under its own weight, but the act of collapsing will heat the core. This process of collapsing and heating continues until the core temperature reaches about 100 million degrees K and the helium ignites.

4. At this temperature, three helium nuclei will fuse into the nucleus of one carbon atom. This fusion releases much more energy than the fusion of hydrogen to helium, which will force the expansion of the Sun.

**E.** As the Sun expands, its surface will cool to about 2000 degrees and it will glow red.

1. It will first grow to fill the orbit of Mercury, then Venus, and then Earth. At this point, the Sun will occupy about half the sky.

2. Even though the Sun's surface temperature is dropping, Earth will heat up. Eventually, our temperature will become the same as the Sun's because we will be orbiting on the surface of the Sun. As we said earlier, the oceans will boil and evaporate, the atmosphere will evaporate, and all life will be vaporized.

**3.** As Earth is engulfed, it will run into the gas that is the material of the Sun, which will slow down its orbit. Earth will spiral down into the Sun and evaporate into a puff of smoke.

**F.** The red giant phase of the Sun will continue until it runs out of helium and is left with a core of carbon.

    **1.** Even higher temperatures are needed to fuse carbon and, again, the Sun will begin to collapse and its temperature will begin to rise. However, the Sun doesn't have enough mass to raise the core temperature high enough to fuse carbon. It stops, then, with a core of carbon.

    **2.** The Sun's outer layers are now so far away that they are only tenuously connected. Eventually, the Sun will dissipate into space. At this point, the Sun is called a *planetary nebula*; its core of dead nuclear material is a *white dwarf*.

**II.** Another way the Earth might be destroyed is in a collision of our galaxy, the Milky Way, with the next closest galaxy, the Andromeda.

**A.** In the 1920s, Edwin Hubble performed calculations using the Doppler effect to determine the speed and direction of other galaxies moving in the sky. He found that the universe is expanding; nearly every galaxy is moving away from ours.

**B.** Some galaxies, however, are moving toward ours, including Andromeda. It is now 2.4 million light years away, but we are on a collision path with it at the rate of 100 kilometers per second, or one-quarter of a million miles per hour.

**C.** It is possible that the Milky Way has some sideways motion that will throw us into orbit around Andromeda, but thus far, most evidence suggests that we are on a plunge orbit down to the center of that galaxy. Because Andromeda has two or three times the mass of the Milky Way, our structure is likely to come out the loser.

**D.** One light year is 5.8 trillion miles; therefore, this collision will not take place for five to seven billion years.

**E.** Computer simulations of such collisions reveal that the galaxies involved become twisted and torn by tidal forces and the force of gravity. Such collisions take several hundred million years and, in fact, we can see pairs of galaxies in every stage of collision to verify the computer models.

**F.** We do not believe that stars will hit each other in our collision with Andromeda, but a star may come close enough to Earth to pull our planet into a different orbit. This orbit could be disastrous if we are too close to, or far away from, our new star.

**G.** Earth might also be flung away from any stars, becoming an *interstellar planet*. Our temperature would drop to the temperature of the universe—three degrees K. Our oceans would freeze and our atmosphere would liquefy. Some life might be able to survive using the geothermal energy beneath the frozen surface, deep in the Earth's crust.

**III.** Finally, the world could end with the heat death of the universe.

**A.** We know that since the Big Bang, the universe is expanding and accelerating—against the will of gravity, which would try to have it collapse back on itself. This accelerating force of the universe is known as *dark energy*.

**B.** As the universe expands, its temperature drops. When the universe was 1/1000 of its current size, its temperature was 3000 degrees, and its temperature now is three degrees. When the universe expands to 1000 times its current size, its temperature will be 1/1000 of three degrees.

**C.** In the distant future, all gas clouds will have made all the stars they can with their supplies. All the stars will burn out. Galaxies will start to turn off or move beyond the visible edge of our horizon. The night sky will grow dark.

**D.** All processes requiring the movement of energy will cease. There will be no earthquakes, hurricanes, or volcanoes. Any such movement creates heat, which will be radiated into frozen space.

**E.** When these processes shut down, the universe will end—not with a bang, but with a whimper.

**Suggested Reading:**

Adams, Fred, and Greg Laughlin. *Five Ages of the Universe*. New York: Free Press, 1999.

Ward, Peter D., and Donald Brownlee. *Rare Earth*. New York: Springer-Verlag, 2000.

Wheeler, J. Craig. *Cosmic Catastrophes*. New York: Cambridge University Press, 2000.

**Questions to Consider:**

1. What happens to planets when two stars come close to each other?

2. What information allows us to deduce that the entire universe will one day reach nearly absolute zero?

# Lecture Five—Transcript
## Ends of the World

Welcome back to My Favorite Universe. In this particular lecture we're going to review things that can bring an end to Earth. The title of this one is "Ends of the World."

Typical references to the end of the world that you might hear might include a rampant virus that totally decimates an entire species, if not some other species, then perhaps humans themselves. There is a lot of talk, especially in the days of the Cold War, about global thermonuclear exchange bringing an end to Earth, or perhaps just the extent to which we are destroying the environment. That's often how people talk about the end of the world. But these "save the Earth" slogans, really they don't mean save the Earth. That's not what they mean. Earth, the planet, is going to be here no matter what we do. It's actually egocentric to say, "Let's save Earth," when you're only really talking about saving *Homo sapiens*. Earth was here before us, it's here now and it's going to be here long after we are gone and a memory in the fossil record.

What I want to talk about are real scenarios that put the entire planet in jeopardy. I can think of three scenarios for this lecture. We actually won't live long enough to see any of them. So why am I talking about them? It's because the formulations of astrophysics allow me to tell you about them, so why not? Maybe, if our species lives long enough, our very distant descendants will care about what these prognostications are. The laws of physics and the phenomenon in the cosmos tell us what is going to happen in the very distant future.

Let's start off with the death of the Sun. Following the death of the Sun, we'll get to the collision between our Milky Way galaxy and our nearest neighbor, the Andromeda galaxy. The third of these will be the heat death of the cosmos.

Let's start with the Sun. We know what the Sun is made of. It's made of gas, hydrogen and helium mostly and some residual of other materials. There are colleagues of mine who specialize in stellar evolution. Give them the birth ingredients of a star and they crank it through a machine. They work through the forces, and the thermonuclear interactions, and the whole life cycle of a star and they produce what are basically actuarial tables for the stars. This one is this old, and it's going to last for another this many years and it's got this kind of health. Oh, a black hole is a little too close to it; it's going to have an early death. They figure this stuff out.

Let's look at the actuarial statements for the Sun. The Sun was born about five billion years ago, giving birth not only to itself, but also to the entire solar system. We date Earth back to about 4.6 billion years. The Sun is going to live another five billion years, so we are exactly mid-way. We are middle aged. We now are referring to the Sun.

How about the Sun's structure? Let's look inside. Starting at the Sun's surface, which is a hot and turbulent place with strong magnetic fields embedded in the plasma, that is the outer surface. These magnetic fields grab onto the structure of the Sun, and they snap and twist, and pieces of the Sun burst forth. It is a busy place. On the surface, it's about 6000 degrees Kelvin (K); the core temperature is about 15 million degrees K. It is hotter as you get toward the center, as you might expect.

In the center is the source of all the energy generated within the sun. We call it the "core," and in the core is where the thermonuclear fusion takes place. It is stable. It is in equilibrium. Hydrogen atoms, the nuclei, are being brought together to make heavier nuclei. If you take four hydrogen atoms, what do you get at the end of that? You get helium. We go from hydrogen to helium, and the act of doing that releases energy because the act of doing that loses mass. We use the $E=mc^2$ equation of Einstein for that, the first equation any of us ever learned, even before we knew what it meant. That equation applies to the center of the Sun. That energy is being generated and it's got to get out. It's trying to get out.

The sun, by the way, wants to collapse under its own force of gravity, but it has a hard time doing it because this energy is working its way out. The balance of forces and pressures between the thermal energy caused by this thermonuclear fusion and the gravitational pressures trying to collapse the star are in balance. The Sun basically keeps its same shape and the same rate of energy output for billions of years. It's a very efficient process.

Before we knew that it was thermonuclear fusion going on in the center of the Sun, people postulated that it might be a lump of coal in the center. That was back in the 19th century when coal was what produced energy in your homes. It was fascinating to bring your home life into the cosmos on the assumption that it's just this big charcoal briquette out there in the center of the solar system.

This party is not going to last forever. The conversion of hydrogen into helium can only continue while there is hydrogen in the core of the Sun at these temperatures. One day the Sun is going to run out of hydrogen in the core. You need these high temperatures for a very simple reason. The nucleus of a hydrogen atom is a single proton. The nucleus of any hydrogen atom is a single proton. I just told you that we're bringing together these protons to create a brand new nucleus, to create a helium nucleus. How am I supposed to bring two protons together when they have the same charge, a plus charge? As you know, like charges repel. Here are these protons and when I try to slam them together, they repel.

The only way you can get them to actually touch is if you create a high enough temperature so that their speeds overcome that repulsion. In high temperature gases, particles are moving very fast, so they overcome. It's like rolling something up a hill and it rolls back. You say, "Let me roll it a little faster." It almost gets to the top, and then it rolls back. Then you really shove it and it goes all the way up to the top, over the cliff, and down the other side. That barrier has to be overcome, and the high temperatures in the center of the star overcome it. They link together and, at the moment they touch, a brand new force takes over. It's got to be a strong force to keep two protons together. In fact, it is a strong force; we call it the *strong force of nature*. You don't have much encounter with that in everyday life, but it's there and it's operating in the center of the Sun.

When the Sun runs out of hydrogen, a big blob of helium will be left in the center of the Sun. What do you do with the helium? Let's try to fuse it into yet a heavier element; but helium won't fuse at 15 million degrees. It's no good—it's not hot enough—because helium has two protons. If you try to bring two protons together against two other protons, that has even more repulsion. You need an even higher temperature. If we had a knob, we'd just say, "Okay, turn up the knob. Let's get to the next temperature, the next magic temperature to fuse helium."

The Sun actually has a built-in mechanism to do that. It's called the *loss of equilibrium*. In that loss of equilibrium, when there's no more hydrogen and the energy snuffs out, there's nothing to support the star. What does the star do? It collapses under its own weight, and the act of collapsing heats the center of the star. It's a marvelous regulatory system inside of these stars. It heats it until it collapses and heats, collapses and heats, down in the core until it ignites helium. That's at about 100 million degrees K in the core, much hotter than what it took to fuse hydrogen to helium.

When you look at the nuclear pathways, three helium nuclei fuse to make the nucleus of one carbon atom. That releases energy, a lot of energy, and much more energy than hydrogen going to helium in total. How do you get that energy out? The star has to respond somehow. As energy is trying to get out, the star begins to expand and expand and expand. That's going to happen to the Sun. Five billion years from now it will run out of hydrogen, it will start fusing helium into carbon after it collapses a little bit and gets the temperature up and starts cranking out the energy. It will start doing that and it's going to start to swell. You will look up in the sky and say, "Hmm. The Sun is getting kind of large!" As it gets large the outer surface will cool, it will become cooler. Astrophysically, when I say something is "cool," it's still thousand of degrees.

The surface of the Sun will drop from 6000 degrees down to about 2000 degrees. It will start glowing red and become enormous. How big? It will first fill the entire orbit of Mercury, the closest planet to the Sun. It will engulf Mercury then fill the orbit of Venus. It will engulf Venus and fill the orbit of Earth. Imagine what that will look like. Just imagine! What would sunrise be? It would be the whole horizon just rising up. Eventually, if it engulfs the entire orbit of Earth, the Sun will occupy half the sky— half the sky.

In an artist's rendition of the scenario, a red giant is depicted about to engulf its next victim. In this case, an attempt has been given to show Earth and how the temperature of the surface of the Earth might be on the rise. As I described, the temperature of the Sun's surface will drop from 6000 down to 2000 or 3000 degrees; but, as that surface gets closer to Earth, the radiant energy that hits Earth will go up. Eventually, the temperature of the Earth will become the same as the temperature of the Sun's surface, because Earth will be orbiting on the surface.

That, once again, will be bad. Because at 3000 degrees, what begins to happen? If you heat the oceans, they will come to a rolling boil and they will evaporate into the atmosphere. The atmosphere itself will get so hot it will evaporate, laying bare the Earth's surface, without protection—without any way to cool itself. Now the surface starts rising in temperature and all known life will be vaporized. Rocks will be rendered aglow, like what goes on inside of a volcano. As Earth is engulfed, it slams into the gas that is the material of the Sun. It's bad for the orbit. The orbit was happy out there in its near vacuum, with nothing to force a decay in its motion. Now there is star material there. Earth will plow through it, it will slow down its orbit, and Earth will descend, spiral down in and evaporate into a puff of smoke. That will happen in about five billion years—the end of the Earth.

I once wrote a planetarium show that depicted just that scenario up on the dome; the Sun getting larger and I had sort of ominous music— unlike anything else I've ever written for that audience. I wrote about Big Bangs, meteor impacts, all kinds of bad stuff I had written about; but, when I wrote about the Sun getting bigger in the sky, I got mail for years from parents who told me their kids could not get to sleep or they woke up with nightmares, because they were worried that that was going to happen next week or in their lifetime. So, maybe there's just something scary about that notion that something we've taken for granted—the stable Sun—would somehow be responsible for our own demise; not only the demise of life on Earth, but of course, of the entire planet.

This red giant phase of the Sun will continue until it runs out of helium and has a core of carbon left over. Now we want to fuse carbon. How do you do that? Will 100 million degrees do that? No, you need a higher temperature. Let's try it. Here you have the sun with its core of carbon, but now you have no energy source. We've been there before. What happens? The interiors of the Sun will begin to collapse. The temperature will begin to rise, but there is not enough mass to raise the core temperature up to the next magic place on the thermometer where you can begin to fuse carbon. There isn't enough mass to get there, so the Sun stops with a core of carbon.

The outer layers will have gotten so far away from the Sun that they will be only tenuously connected to what was formerly their home. Eventually, the Sun dissipates out into space and will make a rather beautiful view when viewed in other solar systems. In fact, we have pictures of such objects. We call them *planetary nebulae*, not because it has much to do with planets, because it is really just part of the life cycle of a star. It's just because when William Herschel first discovered these in the sky with his telescope, a few hundred years ago, he saw they looked like disks on the sky. Just like the way planets looked. Stars are just points of light, but you can see the full surface of a planet. These planetary nebulae were so circular that he said they resembled planets, and he called them "planetary nebulae." There's nothing deep about it; it's really a misnomer and it has stuck.

Notice the core of the star that has been laid bare. That in the center is what we called a *white dwarf.* It's no bigger than the size of the Earth and its dead nuclear material. The rest of that is the outermost layers of the red giant; and this is so large, the full extent of our solar system only comes out just the tiniest bit from that central point of light. By the time it is a full-blown planetary nebula, it is enormous, many, many, many times larger than the full extent of the solar system. There are thousands of these in the night sky, and they are gorgeous. You can make whole catalogs of them. NCG6751 is another one, another planetary nebula. It looks almost explosive, although it did happen rather gracefully. The center is the white dwarf. It's either composed entirely of helium, or in our case, it will be composed of carbon. In any solar system, any system of planets like this one had, they are kissing their days goodbye.

If we are on Earth and that begins to happen to the Sun, that kind of makes interstellar space travel a high priority, doesn't it? While the Sun is on its way out, we at least want to be able to travel between planets. Let's go out to Mars, as the Sun is coming over our back. Maybe Mars will be nice and warm like Earth. If it starts coming out closer to Mars, maybe we'll find an asteroid to hang out on or one of the moons of Jupiter. We've got to be able to do something. We can't just sit here like a chicken in a pot ready to get cooked.

Let's go on to the Milky Way. The Milky Way is our galaxy. It's going to have problems of its own, but these are kind of fun problems. This is not a picture of the Milky Way; it's a Milky Way look-alike. I can't step out of our galaxy, look back, take a snapshot, and go back in and show that to you. This is a look-alike. We've got a little nuclear bulge in the center. We have a spiral shape, and if you saw this edge-on, it would be quite flat. The Milky Way look-alike is one of the countless galaxies in our night sky—just to give you a sense of what we look like. That system is actually the blurred, fuzzy light of 100 billion stars.

Back in the 1920s, Edwin Hubble, the famous American astronomer after whom the Hubble space telescope is named, looked around the night sky, found all the galaxies, did a little Doppler measurement on them—slightly different from, but similar to in principle, what a police radar gun does when it gets the speed of your car on the street—he looked up at all the galaxies in the sky and said, "How fast as they moving?" Are they coming toward us? Are they going away? With the help of others, he made a whole catalog, and he found that nearly every galaxy was moving away from us. That was the discovery of the *expanding universe*.

Some galaxies, however, are moving toward us. We have motion toward each other. One of those galaxies is the Andromeda galaxy, the nearest galaxy to our own. It's not surprising. If all galaxies have random motions, and the expansion of the universe becomes more and more significant the farther out you get, then you might expect the nearest galaxies, with whatever is their speed, to be overcoming the expansion of the universe. In fact, that is just what's happening. These random motions have overcome whatever is the expansion of the cosmos in the nearby universe, and we are on our way toward each other at a speed of about 100 kilometers a second. In ordinary terms, that is about a quarter of a million miles an hour. Fast.

Andromeda, the closest large galaxy to our own, is about 2.4 million light years away. It's beautiful. It has a satellite galaxy in its vicinity orbiting with the rest of the 300 billion stars that are in it. Take a close look; because that is the place we are headed. Strictly speaking, we are approaching each other, and the closing rate is a quarter-million miles per hour. Collision is imminent, assuming that we don't have too much sideways motion. It is possible that we have sideways motion we haven't measured yet that will take us in orbit around it, but the evidence thus far suggests that we are in a plunge orbit right on down into the center. If two galaxies collide—and Andromeda is two or three times the mass of the Milky Way—it's going to be a bad day for the structure of the Milky Way galaxy.

When will that happen? It's 2.4 million light years away and a light year is the distance light travels in one year. It's about 5.8 trillion miles, if you want to know that sort of thing. It's 2.4 million of those away. It sounds far, but we are going to close that gap. It will take between five and seven billion years to do it. If we survive the death of the Sun by whatever means, we'll be around to see the collision between Andromeda and the Milky Way.

We have computer simulations that tell us what this will do. We have a very strong idea of what will happen. The galaxies feel tidal forces between each other where one side of the galaxy is closer than the other so it feels a stronger force of gravity. It gets tugged and twisted and turned. The thing looks like a train wreck when it's done. We didn't make this up. We do the calculations and we say, "This looks like a mess." We look out in the cosmos and we find galaxies that are in the middle of collision. By the way, a collision takes several hundred million years to execute, so we're not sitting here eating our sandwich watching for a collision happen.

There are enough galaxies out there so you can see some that are this far, some that are that far, and another set like that, and another that is intersecting, and others that are all distorted. The universe is big enough so that you find pairs of galaxies in every stage of collision. In that way, we can test our models to see if galaxies were born looking like that or if there is some story that is unfolding. If you compare all these pictures to our models, you can see what each stage looks like. We have the whole evolution of the collision mapped out. It will happen in five to seven billion years.

What actually goes on? Space is pretty empty. We've got big gas clouds and when they collide, they'll stick like hot marshmallows and will give birth to new generations of stars. Stars, however, are so separated that as many stars as are coming into this encounter, they're not going to hit each other. Statistically they're just not going to hit, so that's not going to be a problem. We can sit back and watch.

What I worry about, though, is a star that comes kind of close; not hitting the Sun, but close enough so that maybe we—Earth—lose our gravitational allegiance. What happens if another star comes closer to Earth than the Sun is? Earth will start to respond and react to that force of gravity. In fact, Earth could be stolen from the Sun. It could be like a fly-by looting. If that's the case, fine. Then we'd have a new home star; but maybe Earth is not at the right distance. Maybe it would end up in an orbit too close and all the water would evaporate. Maybe the orbit is too far and all the water would freeze. We need liquid water for life, as we know it. That's not going to be a good thing. It's probably going to be bad for us, but at least we'd have a home star.

In another scenario, Earth—or any other planet—could get flung out from both stars and become an interstellar vagabond. That would be bad, too. What happens there, as an interstellar vagabond? You'd have no home star. No source of energy. Quite the contrary to the Sun becoming a red giant, the oceans freeze. The atmosphere liquefies, and all life as we know it ceases; except, wait a minute there is still some energy. There might be some geothermal energy sources deep down in the Earth's crust. We've got that now. We have life forms that thrive on geo-chemical energy. They have nothing to do with the Sun. It may be that there are thousands of interstellar planets that have life thriving beneath their frozen surface, existing on the energy sources left over from the formation of the planet in the first place.

How cold do we get if we're flowing out into space? Real cold. Without a sun, the temperature drops to the temperature of the whole universe. We can measure the temperature of the whole universe; it's about three degrees on the absolute temperature scale. That is cold. We kind of radiate all of whatever heat we had into space and become this little ice cube. That's a bad day for Earth. You hope that no star comes that close.

Perhaps the most terrifying of all these deaths is the heat death of the universe. The Big Bang is the most successful theory of the universe ever put forth, and it describes an expanding universe. Expanding against the will of gravity that would try to have it collapse back, because we don't have enough gravity to accomplish that. We also know that there is this new pressure that was discovered. We call it *dark energy*. That's the accelerating universe. We know that the universe not only doesn't have enough mass to bring it back, there's a force that will make sure that never happens. It's the accelerating universe. We're on a one-way trip.

As the universe gets bigger, its temperature drops. When we were one one-thousandth (1/1000) our present size, the universe was about 3000 degrees. Today it's three degrees. When the universe gets 1000 times bigger than it is today, then it's one one-thousandth of three degrees. That's cold.

In the distant future all the gas clouds run out, because they will have made all the stars they can with their supplies. All the stars will burn out. When we start running out of our supply, our stars and galaxies start to turn off, one by one. The night sky goes dark. The galaxies that are in our night sky now, whether or not they turn dark, the acceleration of the cosmos will move them so fast that they will go beyond the visible edge of our horizon. They will be shifted out of our view.

Not only that, all processes in the cosmos, anything that relies on energy moving from one place to another—machines, volcanoes, and earthquakes have a build-up and release of energy—all the movement of energy will cease; no earthquakes, hurricanes, volcanoes, none of that. No machines. It all winds down, because every time a machine moves, it makes heat and the heat gets radiated to the frozen space—the frozen temperatures of the universe. Everything that ever could happen in the universe, ceases. The universe is cooling like a pie out on the windowsill, radiating its heat out to the cool air. It's a bad day for the universe. When that happens, we can assert that the universe, though born with a Big Bang, has died, not with a bang, but with a whimper.

# Lecture Six
# Coming Attractions

**Scope:**

Statistically, asteroid impacts represent the biggest threat to human civilization from nature. Most are small and do little damage, but a few are large enough to render the entire species extinct. Evidence in the fossil record tells us that such asteroids have hit before Earth. One layer of rock will show the Earth teeming with life while in the next layer, all life is gone. Such extinction was originally attributed to volcanoes or climate changes, but we now know that about every 100 million years, Earth is hit with a deposit of energy sufficient to cause the loss of 50 to 90 percent of all species. This lecture looks at our risks of getting hit with an asteroid and what we can do to avoid such an occurrence.

## Outline

**I.** You have the same chance of being killed by an asteroid as being killed in a plane crash. How is that statistic true? Calculations tell us that every 100 million years, an asteroid will kill 10 billion people. About 100 people may be killed every year in plane crashes. If you multiply 100 people by 100 million years, you arrive at 10 billion people.

**II.** To examine the catastrophe of an asteroid impact, let's look at the early formation of the solar system.

    **A.** The solar system began as a large gas cloud with the right mass and material to make one star in its center—the Sun.

    **B.** Material was left behind after this formation, including heavy elements and leftover hydrogen and helium. This gas began coalescing and forming molecules. In turn, the molecules began to form into dust, and the dust began to form into rocks. The result was rocky and icy debris in the universe.

    **C.** Some pieces of this debris were larger than others and had more gravity. These pieces grew by accretion of more debris, building larger masses out of scattered smaller objects. When such a mass grew large enough, gravity shaped it into a sphere. Ultimately, these spheres became the planets in our solar system.

**D.** Not all material in the universe ended up as part of the planets. The universe still has a good deal of debris, and in fact, for 600 million years after the formation of Earth, our planet was still being hit. This is known as the *period of heavy bombardment.*

1. During this period, the impacts were so numerous that the surface of the Earth didn't have time to cool down in between. The surface temperature was high enough to kill any life that might have developed.

2. Low levels of iron in the Moon throw light on how it was formed at this time. Computer models postulate that a Mars-sized impactor hit Earth and threw large chunks of the crust into space, which coalesced and formed the Moon. Earth had already segregated its materials, with iron toward the center and lighter rocks on the surface; it makes sense, then, that the Moon has low levels of iron.

3. You usually hear that the span between the formation of Earth and the start of life is 800 million years. The correct figure is four billion years, because during the 600 million–year period of heavy bombardment, Earth was sterile and life could not have developed.

4. Many of the ingredients of early life—carbon, nitrogen, oxygen, and silicon—were brought to Earth's surface by these impacts. In fact, water was brought to the surface by comets.

**E.** Earth continues to accrete, but the impact rate has become more manageable. We have had enough time between impacts for species to evolve.

**III.** What are the leftovers from formation of the solar system?

**A.** The universe has tens of thousands of asteroids in an asteroid belt. This is a region of the solar system in orbit around the Sun between Mars and Jupiter.

**B.** The universe also has trillions of comets, some near and some far. Comets can be viewed as fossils of the early solar system.

**C.** Most comets and asteroids don't cross Earth's orbit, but some do. We have avoided a major collision for a while simply by being in the right place at the right time.

**D.** We have identified some comets and asteroids that would do serious damage if they collided with Earth, such as asteroid Eros. Some of these have come fairly close—between Earth and the orbit of the Moon—and we don't usually see them until it would be too late to take action.

**IV.** We have a good deal of evidence that these damaging impacts have hit Earth.

    **A.** For example, geologist Eugene Shoemaker determined that Meteor Crater in Arizona was formed by a meteor impact.

    **B.** Of course, the most well known impact is the one that resulted in the extinction of the dinosaurs.

        **1.** This impact event is called the *Cretaceous-Tertiary (KT) boundary.* The boundary is the layer in rock that tells us what happened before and what happened after the event.

        **2.** After the KT boundary, 70 percent of land species became extinct.

        **3.** The impactor in this event was 10 to 20 miles in diameter and hit the tip of the Yucatan Peninsula 65 million years ago.

        **4.** Modern-day mammals owe their origins to this event. The dinosaurs were wiped out, but small mammals survived and had a fresh ecosystem to populate. Without this planet-wide catastrophe, we would not exist today.

**V.** What is the force of these collisions?

    **A.** Asteroids hit the Earth at 45,000 miles an hour.

    **B.** Some asteroids explode in the atmosphere, creating shock waves that would flatten houses.

    **C.** A large asteroid that hits the Earth's crust will make a crater and spew dirt into the upper atmosphere, cloaking Earth with a darkened sky and preventing sunlight from reaching the base of the food chain.

    **D.** The asteroid that took out the dinosaurs had an energy of 100 million megatons of TNT, the equivalent of five billion atom bombs.

**E.** About every 100 years, we are hit with an asteroid that is about 30 meters in diameter and has the energy of two megatons of TNT, or the equivalent of 100 atom bombs.

**F.** One such event occurred in Siberia in 1908; 1000 square miles of forest was felled and 100 square miles of vegetation was vaporized, but no crater was found. Computer simulation predicts that this destruction was the result of a stony meteorite, about 60 meters across, which exploded in mid-air. Its shock wave and radiant heat killed the forest.

**G.** We determine the rate of impact by looking at cratering history on the Moon, which experiences no weather and no erosion and, therefore, preserves all its craters.

**H.** Other planets are also at risk. For example, Jupiter was slammed in 1994 by a train of about two-dozen comets that had originally been one comet. Each one of these impacts had enough energy to have caused the extinction of the dinosaurs on Earth.

**VI.** What can we do to avoid getting hit by a large asteroid?

**A.** One approach would be to blow it out of the sky, but a direct hit would be necessary.

**B.** We might also send up rockets to attach to such an asteroid and nudge it out of the way. If we could predict the collision 200 years in advance, we would need to nudge the asteroid only one centimeter a second to avoid impact.

**C.** We need to catalogue every asteroid that is bigger than about a half mile across. A collision with something of this size could change Earth's ecosystem.

**D.** We are the only species in the universe that has the technology and the knowledge to protect ourselves, but if we don't take action, we will go the way of the dinosaurs.

**Suggested Reading:**

Lewis, John S. *Mining the Sky*. Reading, MA: Addison-Wesley, 1996.

———. *Rain of Iron and Ice*. Reading, MA: Addison-Wesley, 1996.

Vershuur, Gerrit L. *Impact: The Threat of Comets and Asteroids*. Oxford: Oxford University Press, 1997.

**Questions to Consider:**

1. How does the probability of being killed by an asteroid compare with the probability of dying in a plane crash?

2. If an asteroid or comet hit Earth, how big would it have to be to wipe out civilization, as we know it?

# Lecture Six—Transcript
## Coming Attractions

Welcome back to My Favorite Universe. I'm going to talk about a subject that is on a lot of people's minds, a subject that has been worked into two feature-length Hollywood motion pictures. I'm talking about killer asteroids; asteroids that are out there in space, normally minding their own business, and occasionally a wayward one comes our way and slams into Earth. If you look at the statistics of asteroid impacts, they represent the biggest threat to human civilization that exists in nature. While most are small and cause little or no damage, the fact that there are a few that could render the entire species extinct, means we ought to be paying attention to these things.

We know Earth has been slammed before by some combination of comets and asteroids, because there's evidence in the fossil record. Paleontologists have known for a long time that there are periods in Earth's history where it was teeming with life, then they hit some boundary in the sedimentary rock, and then most life is gone in the next layers. Before we had a deep understanding of impacts, it was often ascribed to too many volcanoes or a climate change or the Ice Age or whatever. We now know, every now and then, Earth gets a deposit of energy sufficient enough to cause the damage observed in those layers.

We know that in that fossil record, about every hundred million years—plus or minus a few—Earth loses anywhere between 50 to 90 percent of all the species on the planet. It's more than just, there's life here and then no life on the layer that follows it; maybe all the life migrated away from that spot. If you look all around the world, the evidence is irrefutable and incontrovertible, that when the life disappeared here, it also disappeared on the other side of the Earth. Nobody was running from one place to another. Something happened catastrophically.

We're going to spend the next few minutes talking about what this risk is and trying to get closer to it. To come to an understanding of what the risk is and is there anything we can do about it, being intelligent human beings—the only intelligent species ever to walk the face of the Earth. I'll define intelligence by the capacity to maybe build something that will knock it out of the sky.

Let's begin with a statistic. Perhaps you don't know, but the chances that your tombstone will read, "Killed by an asteroid," are the same as the chances of your tombstone reading, "Killed in a plane crash." You might say, "How is that possible?" You know of people, either first hand or second hand, who might have died in a plane crash. I'm sure you don't know anyone who's been killed by an asteroid. A couple hundred people a year die in airplane accidents; zero die a year from asteroids. What am I saying here? It's because you have to look at it, at how it works. We'll go 100 million years, and then an asteroid large enough will strike the Earth and kill 10 billion people. So that's 100 million years, 10 billion people. Now let's talk about airplanes. Airplanes, on average, might kill 100 people a year; but, if it does that for 100 million years—at the end of that same baseline of time—airplanes have killed 10 billion people. That's why those two data are equivalent. The risk of death is the same in an airplane as it is by asteroid.

This is something we should worry about. It's something we need to alert the legislature about. The problem is, of course, to go to someone who needs to get reelected every two years with something that might hit in the next 100 million—doesn't square with the attention span of Congress. I'd like to believe that, in the next few minutes together, I will convince you of why it's an important step to take.

Let's look back at what started all this: the early formation of the solar system. We started out as a large gas cloud with the right mass to make one star in its center, the Sun. The rest of the material, enriched with heavy elements—elements beyond hydrogen and helium, the kinds of elements that make planets—mixed with whatever hydrogen and helium that did not participate in the formation of the star. We have gas coalescing and making molecules, molecules making dust, dust making rocks, and also volatile dust making ices. The result is rocky debris and icy debris, depending upon the distance from the Sun where it formed. If it's too close, the ice won't condense out. Most of our comets reside very far from the Sun, for example.

There is all this debris. Some debris is a little bigger than the others, so that will have a little extra gravity, a little bigger cross-section to impact. They'll grow by accreting more of the debris. Through this method you can actually build and create large masses out of what was previously just this scatterbox of small objects in orbit around the host star. By these collisions, small objects become large. Slightly bigger ones grow bigger faster than slightly smaller ones. When they reach a large enough size, gravity then shapes them into a sphere. We spent a whole lecture on that, on being round. This ensemble of spheres is what we now describe as our solar system: the planets and the planet's moons.

Not everything ended up as part of the planets. There is still a lot of debris out there. In fact, even after the planets were formed, there was a period of sort of "clean up" after the initial formation. There were 600 million years where planets, fully formed, were still hit. Technically, they were accreting, but they no longer were growing appreciably in their total mass. They were simply being subjected to impacts. We call this 600-million-year period the *period of heavy bombardment.*

Impacts came at such a rate that there wasn't time for the surface to cool down, between this impact and the next one. The average temperature of the surface remained high. It was high enough to kill anything that might have formed, high enough to break apart complex molecules, high enough to consider the surface of the Earth a sterile environment.

The Moon, if you look at it, for an orb that size in the solar system, it ought to have much more iron in it than it does. The Moon has very little amounts of iron; why is that so? The latest models show that the Moon was spawn—was hewn out of the surface of the Earth. That a Mars-size impactor had a glancing blow onto Earth's crust, grabbing whole pieces of the Earth and thrusting them back out into space. They coalesced again, forming a sphere in orbit around the Earth, making today what we call the Moon. Since the Earth had already segregated its materials, with iron being toward the center and the lighter rocks toward the surface, it makes sense then that the entire orb of the Moon would be made of just the rocky material with very little iron, because the iron had already been filtered out, so to speak.

Think about that. When did life begin? The earliest records of life fossils on Earth go back about 3.8 billion years. That's the earliest fossils. And how old is Earth? Earth is 4.6 billion years. If you subtract those two numbers you get 800 million years. All right, it took life on Earth 800 million years to start. That's a commonly quoted number. But don't re-quote that number, because that's not fair to the fertility of organic chemistry, because you can't include the 600-million years that we were sterile. Nothing could have formed. That's just not fair. Wait for that to die down a bit, so that at least between impacts, Earth had a chance to cool and make complex molecules. Take 4.6 billion, minus 600 million, and you're right back to four billion years. Start your "life clock" at four billion years ago. Wait around. Where's life? Wait 200 million years, and there are your one-celled creatures.

Life was remarkably fertile. The early ingredients were carbon, nitrogen, oxygen, and silicon—the ingredients of life—many of which were brought to the surface of the Earth by those impacts. Water was brought to the surface by comets. This period of heavy bombardment was kind of like "soup starter" for life in the early Earth.

Earth continues to accrete, but not at that rate, and that accretion—that's the official word, "accretion." I actually don't like that word, but it's the official term. People more formally use the word "accretion" when they really should talk about species-killing—ecosystem-destroying impact—because that's what these things are. And they're the leftovers of the solar system. They're still out there. It's just that the impact rate has dropped into what we might consider a manageable rate. There is enough time between really bad impacts to evolve species from one form to another. In that respect, it's okay.

What are the leftovers? First, we have tens of thousands of asteroids in an asteroid belt. That is a region of the solar system in orbit around the Sun between Mars and Jupiter. We have trillions of comets. Some are nearby, like comet Halley, which has only a 76-year period. Other comets go very far away and have orbital periods of thousands of years and tens of thousands of years, but if you add them all up, there are trillions of them. Since these are left over from the early solar system, if you capture one, or experiment on one, or get a piece of one, they are like fossils of the formation of the solar system. Anytime you see someone trying to send a mission to a comet, pay attention to that, because these comets have been hanging out there for billions of years, untouched, unscathed, unaltered. I don't know about you, but every time I drink a cup of water I think about comets as the delivery system for this refreshment. It keeps me connected to the cosmos.

Most comets do not cross Earth's orbit. Most asteroids don't cross Earth's orbit. We don't have to worry about those because the dynamics of the system prevent them from ever crossing Earth's orbit. But, there are samples of comets and samples of asteroids that do cross Earth's orbit. Those are the ones we're worried about, and they cross every time they come around. Why don't they hit us if they cross our orbit? You've crossed the street before, the same street that trucks have gone by, and you didn't have an encounter with the truck, because you crossed at different times. These are very big orbits. There's a lot of real estate out there, so you can go a long time without ever actually colliding. You're at the right place at the right time rather than the wrong place at the wrong time.

What we want to do is actually catalog the bad ones. Let's look at a real asteroid. I've got one, asteroid Eros. It's a potato-looking thing, if I've ever seen one—a funny-looking potato. That's a real asteroid called Eros, which would do serious damage if it hit the earth. Every now and then we see one of these come by in a close approach. A close approach might be between Earth and the orbit of the Moon. Even if it's twice the orbit of the Moon, that's close. People call it a near miss, and I call it a near hit. It didn't nearly miss us; it nearly hit us. If it nearly missed us, it would have hit us. I think of them as shots across our bow. We see them when it's too late to do anything about it, either coming toward us or receding from us.

We've got to watch out. A few years ago there was an asteroid whose orbit we calculated to intersect with Earth sometime in the next 100 years. The original calculation gave a high probability of that. Everyone went berserk; when I say everyone, I mean the media went berserk. Every astronomer in every town was shoved in front of a TV camera and asked, "Is this bad? What's it going to do?" Astronomers became the most important people in the world for 48 hours.

What happened? Better data became available in that short amount of time. We recalculated the orbit and found that no, it's not going to collide with Earth, but it wasn't before the press gave it enormous levels of attention. My favorite headline of them all was from the *New York Post* the day after the calculation was refined. Instead of saying, "The asteroid almost hit us, but it's not any more," what headline did they give us but, "Kiss Your Asteroid Goodbye." I just thought that was a brilliant headline in the style that only the *New*

*York Post* can put together. I remain astonished by the auxiliary headlines that go together with these major other headlines. "Kiss Your Asteroid Goodbye," this is an article about what might have been the end of all civilization, and it's combined with a movie review of Leonardo DiCaprio's recent film. This is the world in which we live.

Do we need evidence that we've been hit other than extinction layers? We've got evidence. Have you ever been to Arizona? Arizona is known for its holes in the ground. The one most people talk about is, of course, the Grand Canyon; but the one I think about most is a big crater, Meteor Crater, Arizona. Before it was called Meteor Crater, it was called Barringer Crater—because people didn't know that a meteor made it. It took the efforts of the geologist, Eugene Shoemaker, who, as part of his PhD thesis, analyzed the structures of the crater and the surrounding areas. He was like a planetary geologist. He knew about craters on the moon and what could cause them. He postulated that this was made by an impact, as opposed to what was previously believed, that it was a volcanic caldera—even though there is no evidence of volcanic activity anywhere for hundreds of miles in that location in Arizona.

That crater is one of the most famous on Earth because of how well preserved it is. It's not heavily eroded. Craters don't keep their shapes very long on Earth because of the weather. The elements are devastating to surface features. To get a sense of how big this crater is, if you played a football game in the center of that crater, there's enough seating for two million people around the crater edge. On my first visit to Arizona I took a trip to the Grand Canyon, and I was under whelmed by it. It was a big hole in the ground and it took however long nature took to make it—I don't know, millions of years, whatever. I went to Meteor Crater, came up to it and that's impressive. I said, "Look at this hole in the ground." How long did it take to make it? One-tenth of a second! I remain impressed with the power of nature simply by my memory of visiting that crater. I have a piece of the larger meteorite that made that crater. It's an iron and nickel meteorite, 90 percent iron and 10 percent nickel. It's very heavy, very dense. You can get a sense of that just by the sound of my putting it on the table.

The most celebrated extinction event in history is when the dinosaurs kicked. When was that? We call that the *Cretaceous-Tertiary boundary*, the *KT boundary*, which is a boundary in the rocks that sets up a difference between what happened before and what happened after. We've always known about that boundary, but no one really knew what caused it. Was that event as significant as Meteor Crater in Arizona? Meteor Crater in Arizona, in fact, is small. It would have wreaked havoc in its local area 50,000 years ago when it hit; but beyond that, it's a minor, minor encounter with an asteroid.

If we're going to blame an asteroid for the KT boundary, it's going to have to be something bigger than that. The paleontologists were in denial. If you're a paleontologist you look for fossils, so you're looking on the ground. They never looked up. They look for solutions to their problems on the ground. Astronomers looked up. "Hey guys we've got stuff coming. I can deliver you significant sources of energy that can take out your dinosaurs at about the rate that all your fossil records are going away." It wasn't until recently when that community finally warmed up to this fact.

Let's take a look at an artist's rendering of what this might have been. With the KT boundary, as I said, 70 percent of all Earth's land species went extinct. No animal bigger than a breadbox survived. Evidence indicates that the impactor was about anywhere between 10 and 20 miles in diameter. It came in and hit the tip of the Yucatan Peninsula. It's known as the Chicxulub Crater. There's a crater there; we found a crater. There's the smoking gun. After hypothesizing an asteroid took out the dinosaurs, we found a crater of about the right size and we dated the crater 65 million years ago. Something bad happened on Earth 65 million years ago. By the way, the dinosaurs went extinct.

Here's another view, showing some pterodactyls flying by. This asteroid hits the ocean in this image. In either case, you're hosed. Don't try to pick your spot; it won't matter. You're dead. I like to think of *T. rex* observing this and then, of course, in the next scene we have of *T. rex* and he's all bones in the museum.

It makes you wonder, though, can we have our cake and eat it, too? Can we say, "Oh, these impacts are bad? I don't like these impacts. They'll kill us." Well wait a minute. Our ancestors, our mammalian ancestors, tree shrews running underfoot at the KT boundary, survived that impact. They spent their whole life wanting not to be eaten by *T. rex*; then *T. rex* gets knocked off and they now have a fresh ecosystem in which to populate. Modern day mammals owe the origin of their diversity, and their spreading across the world, by virtue of the dinosaurs having been taken out of the picture. We can't simultaneously be happy that we live on a planet, be happy that the planet is chemically rich—brought to us by impacts—happy that we're not dinosaurs, but yet somehow resent the possibility of us being knocked out by a planet-wide catastrophe. A planet-wide catastrophe enabled us to be here in the first place. Maybe we should just take it in stride.

What kind of collisions can you have? Think about it. If this meteorite fell from 12 inches above my head and hit my head, it would probably crack my skull. Imagine this going 45,000 miles an hour. It would swat me like a bug. Like a chain mail glove hitting a bug. Some asteroids explode in the atmosphere, and by exploding in the atmosphere they create shock waves. A plane moving faster than the speed of sound leaves a shock wave. We call it a sonic boom. It might rattle your dishes, but a shock wave left by one of these flattens your house. It's a whole other scale.

What happens if it survives the atmosphere? If it impacts Earth's crust, it will make a crater, as we know, but if it's a big enough crater, it will thrust the dirt into the upper atmosphere, into the stratosphere. It will wrap around Earth. It will cloak Earth with a darkened sky, preventing sunlight from reaching the base of the food chain. Even if you were not there at ground zero of the impact—where you would then be vaporized instantly—if you were living on the other side of the world, this stuff affects you because it completely changes your climate.

I've got a table here just to get a sense of how often these bad things happen. We get certain encounters with these stones once a month, once a year, once a decade, once a century; and the rarer the impact, the bigger the impact is. I put this up just so that you know that about every hundred million years there is an asteroid, 10 kilometers in diameter, which has the energy of 100 million megatons of TNT. How many A-bombs is that equal to? It's equal to five billion A-bombs. Can you even imagine that? That's what took out the dinosaurs—that's bad. Let's look at once a century, coming up half way on the table. Once a century an asteroid about 30 meters in diameter has an impact of two megatons of TNT, or the equivalent to about 100 A-bombs. Once a century! Once a century that has happened.

How did we calculate that? There are a couple of famous impacts that we can refer to. One of them is Tunguska, Siberia. At Tunguska, Siberia in 1908, 1000 square miles of forest was felled and 100 square miles of vegetation was vaporized; but there was no crater. What did we learn about it? We studied the event using modern computing and found that a stony meteorite, 60 meters across—here we have one 30 meters, in the same ballpark—it came in and never survived Earth's atmosphere. It exploded in mid-air. The energy's got to go somewhere. There was a blast wave, shock wave, a radiant heat, and that destroyed the forest. You get that every hundred years. If that happened over a city, that would be the end of the city.

Another famous impact, of course, was the KT boundary. All the numbers square with that. How much impact energy the object had that made that crater is the right amount of energy it would take to knock out all the major life forms on Earth. It all matches. We get this rate of impacts from looking at the cratering history on the Moon. The Moon preserves all its craters. There's no weather on the Moon, nothing to erode any features that it has. There are craters, on top of craters, on top of craters. If you analyze that, and look at how often big ones come, how often small ones come, you can produce a table like this one.

Other planets are at risk as well. We're not alone in this. The Moon, Mars, Venus and Jupiter are at risk. Just a few years ago, in 1994, the comet Shoemaker-Levy slammed Jupiter. Those are the co-discoverers. In fact it was the husband and wife team, Eugene and Carolyn Shoemaker along with David Levy, patron saint of amateur astronomers—he's got something like 20 comets. I've lost count of how many comets he has discovered. The fact that that one was named Shoemaker-Levy 9, meant that's the ninth one that comet-hunting team found. They found a comet whose trajectory was going to intersect Jupiter itself; it was going to collide with Jupiter. That comet, formerly in one piece, got broken apart into a train of mini-comets. If you take one comet and break it into two, you've got two comets. It's like cutting up earthworms.

Calculations indicated we were going to see a collision of the first rank, something we had never seen in the era of the telescope. Sure enough, the prediction was made, and in July of 1994, this whole train of comets slammed into Jupiter. Each one of these blobs would have been enough to have taken out the dinosaurs on Earth, and there were two dozen of them running like a train—like lemmings off a cliff—plunging into Jupiter's atmosphere. Whatever effect it had on Jupiter, you know Jupiter has no dinosaurs today. That was a bad day on Jupiter.

I can ask you this, "What are we going to do about it?" If we see one headed our way that has own name on it, what do you want to do? You can play macho man and say, "Let's blow the thing out of the sky." But these are big hunks of rock. You'd have to aim right—if here's the asteroid, you can't explode something here and expect it to affect your target, because there's no air to propagate a blast wave. You need the change in air pressure in order to actually destroy a target, so you actually have to hit it.

There are kinder, gentler ways to accomplish the same thing. One of them is to send little rockets up to it that attach to it and gently sort of nudge it out of harm's way. If you get it early enough in its orbit, you just have to move it by about one centimeter per second. If the collision course that you calculated was for 200 years from now, and you move it now one centimeter per second, 200 years from now it's way off course, just where you want it to be, rather than on course ready to hit the Earth.

Fortunately, the bigger the asteroids are the easier they are to find. Those are the ones that do the most damage. What we really want to do is catalog every asteroid bigger than about a kilometer, about half a mile. Those are the ones that would change Earth's ecosystem. You want to know about all those; they would change civilization. Smaller than that it could be really bad, but we'll recover. Those at a half-mile are the ones we want to track.

An interesting political ramification of this, of course, is that if we find an asteroid headed our way, the asteroid becomes the enemy to everyone on Earth, and all those who were formerly enemies of each other must come together to fight the common enemy. You'd be surprised how quickly you can make friends when you both have an enemy, no matter what your previous relationship was.

I will conclude with the following thought. I don't want to be the laughingstock of the galaxy. We are human beings with a brain. We've got a space program. We know how to launch rockets. I don't want to be caught sitting here watching our species go extinct, being the first life form in the galaxy to have the technology and the intelligence to protect ourselves, yet not actually do anything about it. If we don't do anything about it, we are no better off than the proverbially pea-brain dinosaurs who, in fact, had no choice.

# Lecture Seven
## Onward to the Edge

**Scope:**

In this lecture, we take a break from the death and destruction of asteroids and the end of the universe to wonder at the enormity of the cosmos and question our place in it. Since the 1960s, people have been inspired and uplifted by images brought back from space. These same images have deepened our conception of the universe, bringing us to the realization that Earth is just one world among worlds unnumbered. This lecture takes us to the limit of our vision and asks what questions remain once we reach those limits.

## Outline

**I.** One of the first images brought back from space in the 1960s was a black-and-white photo of "Spaceship Earth."

    **A.** For the first time, we could see large-scale weather patterns from the perspective of space.

    **B.** People also viewed continents without the political boundaries that are delineated on the globe.

**II.** Another image that is equally inspiring is called "Earth Rise," taken by the astronauts of *Apollo 8* in orbit around the Moon.

    **A.** This image increased our awareness of our neighbors in space. We came to realize that the Moon has a sky and a horizon and experiences Earth rise, just as Earth experiences the sunrise.

    **B.** The fact that *Apollo 8* did not land on the Moon enabled the astronauts to take this full-color image of "Earth Rise," the most recognizable photograph ever taken.

    **C.** This image seems more compelling than the earlier black-and-white photos because it is in color. We can see that the Earth is not featureless; it has water, weather patterns, and so on.

**III.** In the early 1980s, spacecraft that had been launched in the 1970s finally reached their destinations and began sending back images. These spacecraft included *Pioneer 10* and *11* and *Voyager 1* and *2*.

**A.** The images from *Voyager* enabled us to see rich detail on the surface of Jupiter. We learned that Jupiter has extreme weather, including a storm system called the *great red spot* that has been raging for at least 300 years.

**B.** We also learned that a great deal of terrestrial action takes place on the moons of planets in our solar system. Jupiter's moon Io, for example, has an extremely active volcano. A sheet of ice that reveals changing patterns of fractures covers another moon of Jupiter, Europa. There may be liquid oceans beneath this ice and even life.

**C.** These images changed our view of the planets as abstractions from an astronomy textbook to the concrete reality of an existence similar to Earth's. These planets and moons are worlds in themselves, just as Earth is a world.

**D.** Such realizations also changed our outlook on our existence. For example, we note that Venus has a runaway greenhouse effect. Its surface is 900 degrees F, hot enough to bake a 16-inch pizza in nine seconds. Is Earth heading toward a similar environment?

**IV.** To me, the most profound and thought-provoking image is one brought back from the Hubble space telescope in the 1990s.

**A.** The Hubble is the first orbiting observatory specializing in visible light. It was launched by the space shuttle and can be repaired in space from the shuttle.

**B.** The fact that the Hubble is in orbit above Earth enables us to see the universe without the turbulent effects of the atmosphere.

  **1.** As light comes from the depths of space, it travels in a perfectly straight path, but when it hits the atmosphere, variations in temperature and density cause that light path to be refracted and dispersed. The result is photos that are blurry and unable to reveal fine detail.

  **2.** In contrast, the Hubble is able to send back very high-resolution photos.

**C.** Astrophysicists must apply for observing time on the Hubble, and as you might expect, many more applications are submitted than are accepted. As part of the director's discretionary time, a period of ten consecutive days of exposures was allocated to document the phenomena in a random and unremarkable portion of the sky. The result of that documentation is the image known as "Deep Field."

1. The portion of the sky that was selected for this experiment was away from the plane of the Milky Way and in a direction where Earth would not block its view. It was also selected to avoid known clusters of galaxies and contain no stars.

2. The patch of sky selected and photographed by the Hubble telescope is near the Big Dipper, but it is extremely small—one one-hundredth (1/100) of the area filled by the full Moon in the night sky. You might think of this area as equivalent to the size of Lincoln's eye on a penny held at arm's length.

3. A total of 342 photos were taken of this identical portion of sky then added together to make one high-quality image.

D. Amazingly, this one tiny patch of sky reveals thousands of galaxies. Practically every spot of light in the photo is an entire galaxy much like the Milky Way, containing hundreds of billions of stars each.

E. Given the precision of the Hubble's imagery, we can get close to the picture and actually look into the galaxies.

1. We see the structures of galaxies, some with spiral arms, regions of extra star formation, and different colors caused by different colors in the stars.

2. In similar photos taken by ground-based telescopes, these galaxies are just smudges.

F. The farthest known galaxy in all the catalogues is seen in this image as a red dot. Why is it red?

1. The wavelength of light is stretched as it travels in an expanding universe. If light is stretched, it becomes lower in energy, and red light has less energy than blue light.

2. We see this distant galaxy not as it is now, but as it was 13 billion years ago. It is one of the earliest galaxies known and serves as a signpost for what the universe was like in the distant past.

**V.** Can we extrapolate from the Hubble "Deep Field" image? If every patch of sky is similar, how many galaxies are in existence?

    **A.** The answer is 50 billion. And each one of those galaxies contains 100 billion stars. The sheer magnitude and diversity of the cosmos give us pause.

    **B.** Just as pictures of the planets changed our abstract conceptions of Earth and the solar system, the Hubble "Deep Field" transformed our vision of galaxies.

    **C.** Researchers now train other kinds of telescopes on that field to add information to the picture. For example, the Chandra x-ray telescope reveals the presence of supermassive black holes in the centers of these galaxies.

    **D.** We have also looked in the exact opposite direction of the Hubble Deep Field to ensure that the original spot is representative. The resulting images of the Hubble Deep Field South reveal a statistical similarity. This new portion of the sky doesn't look exactly the same as the original, but it has the same number of galaxies, the same distribution of colors, and the same array of shapes and sizes.

**VI.** Our new understanding of the cosmos poses deeper questions.

    **A.** Are the laws of physics the same everywhere and through all time? Are there as-yet undiscovered laws that will grant us greater insight into our world and the unnumbered worlds in the universe?

    **B.** Do the planets among the stars in those billions of galaxies have life? Do they have intelligent life? Is that life looking at us in the same way that we are looking at it, or are most other life forms engaged in the quest for food, shelter, and sex, as we are on Earth most of the time?

    **C.** The image of the Hubble Deep Field represents an intellectual journey. As we look at this photo, we have gone beyond stargazing to galaxy gazing—through time and to the edge of the cosmos.

## Suggested Reading:

Voit, Mark. *Hubble Space Telescope: New Views of the Universe.* New York: Harry N. Abrams, 2000.

http://heritage.stsci.edu. The Hubble Space Telescope on-line archives.

## Questions to Consider:

1. What is the primary reason why the Hubble Space Telescope is so valuable to our understanding of the universe?

2. Observing the universe is sometimes said to be like observing though a time machine. What property of the universe makes this so?

*Image Credit: The Hubble Deep Field video, produced by Kenneth M. Lanzetta. Raw images courtesy of NASA and the Space Telescope Science Institute. Partially funded by the National Science Foundation.*

# Lecture Seven—Transcript
## Onward to the Edge

Welcome back to My Favorite Universe. We spent several lectures talking about death and destruction by virtue of black holes or meteor impacts and the like. Let's take a pause from that and talk about life. What it is to be alive and what influences around us force us to take pause in our daily routine and ask, "What is our place in the universe." When I think of influences that have forced me to take pause, being an astrophysicist I'm prone to select images of the cosmos that do this, but I'm not alone in this regard. Many people— you don't have to be an astrophysicist—have been inspired, uplifted, and influenced in very positive ways by images that have been brought back from space, either by astronauts themselves or by telescopes placed in orbit. That forces us to ask the question, "What are the limits of our vision?" Once we've reached those limits, what influence does it have back on our dreams?

One of the first of these powerful photographs was one taken of the Earth. It was taken back in the 1960s and it was not a full Earth, it's a *gibbous* Earth, but Earth nonetheless. Earth seen from space enabled you to see clouds, large-scale weather patterns, for the first time. People looked upon the continents of the world and did not see political boundaries so carefully etched on globes and maps that we buy and we learn about in civics class. The cosmic perspective is different than the political perspective. That had an impact. At around that time in the 1960s, people viewed us as "Spaceship Earth." I remember distinctly the introduction of that as a phrase into common parlance, "Spaceship Earth." Before that image, no one thought of us as that. Here we are together, this fragile little world in orbit around the Sun, like a spaceship.

This image was followed by another image that, for me was equally as inspiring. An astronaut did not take this next one. An unmanned probe in orbit around the Moon took it. It is "Earthrise on the Moon," —in all its black and white splendor. Black and white pictures precede color pictures, so they give you a sense of history, sort of encoded just in the fact that they're black and white. The lunar landscape is in the foreground and Earth is in the background. The Moon has a sky; it has a horizon. It's got Earth in the sky, just the way from Earth we have the Moon in the sky. It's another world.

The fact that Earth, in our own minds, was not the only world in town; the Moon was another world, a place that we might visit.

In fact, shortly thereafter, *Apollo 8* became the first astronauts to leave Earth's orbit and venture to the Moon. They did not land on the Moon, but they went in orbit around it; and the act of going into orbit afforded them the opportunity to watch Earth rise from the Moon. The full-color image, known simply as "Earth Rise," is the most recognizable photograph ever taken. For me, this image is that much more compelling, that much more emotionally rich than the first image of Earth just from slightly high up; because now we see the Earth in color and it's not just some featureless orb out there. It is richly blue. By contrast, we're reminded that the surface of the Moon has no water and it has no atmosphere, it has no weather, and it has no clouds. There's Earth in the distance, this fragile marble, almost aglow. This image forced me to take pause, not simply as an astrophysicist, but as a human being and as a citizen of planet Earth.

This march continued—in fact, the title of this lecture is "Onward to the Edge." I don't know where the edge is, but we're moving onward toward it. In the act of moving onward, we pass signposts. One of them was first leaving Earth. The next one was hanging out on the Moon. What's next? The planets. We haven't sent people to the planets, but we've sent our robotic emissaries. I'm okay with that, because they can take photographs just as good as an astronaut can and beam them back. You don't even have to feed the robot; you don't even have to bring the robot back. It's a very affordable way to do your exploring.

In the late 1970s, early 1980s, spacecraft that had been launched in the 1970s finally reached their destinations, the *Pioneer 10* and *11* space probes and *Voyager 1* and *2* space probes. Voyager, especially, had close encounters with Jupiter, enabling you to see the rich detail across Jupiter's surface. You thought Earth had weather; Jupiter's got weather! It has weather bands and turbulent eddies. It has a storm system; the *great red spot* we call it—because it's a big red spot on Jupiter's surface. That's a turbulent storm that has been raging for at least 300 years—from when it was discovered 300 years ago. We passed by Saturn. We saw the cloud cover on Saturn, too. We saw the rich texture on the surface of these orbs. All of a sudden the planets were not just some other things out there; they, too, became worlds in our conscious.

Up until then we knew planets had satellites, sure, and moons. We knew that, but they were just little dots of light. If you look through the eyepiece of as telescope, it's a dot of light. When we went there we found out that all the action was among the moons. The terrestrial action: Jupiter's moon, Io, the closest moon in orbit around Jupiter, has a volcano on it. It's an active volcano. In fact, it's the most active volcano in the entire solar system. It's not on a planet; it's on a moon of Jupiter.

There's another moon, Europa, a moon of Jupiter. By the way, Galileo had discovered these moons. Out of respect for his efforts in this regard, we call a set of four moons the "Galilean satellites" of Jupiter. If you look at Europa, it has a sheet of ice covering its surface. If you look there now, there are certain patterns of fractures on the ice; if you look there in a month, the patterns have changed. This tells us that maybe there are liquid oceans beneath the ice on its surface. These are worlds. Maybe there's life there.

These are no longer abstractions of an astronomy textbook. They have become extensions of our backyard. They have changed how we think about Earth, simply by recognizing that we're not the only world to think about. It may be, for example, that on Venus, which has a runaway greenhouse effect—it's 900 degrees Fahrenheit on Venus, hot enough to bake a 16-inch pizza in seconds—something went wrong on Venus. We've got a greenhouse effect going on Earth and we've got knobs we're turning, not knowing what the effect will be, the long-term effect. I don't want Earth to look like Venus. Venus is a world; a world that I can now think about in relationship back to Earth. These images of these planets force me, once again, to take pause at our place in the cosmos, not only for me as an individual, but also for Earth. Given the scale of the galaxy and of the universe itself, these are worlds unnumbered.

In the 1990s another image came by, an image brought to us by the Hubble space telescope. The Hubble telescope, let me remind you, was the world's first orbiting observatory specializing in visible light, ordinary light that our eyes are sensitive to. It's an ordinary telescope that has been endowed with the capacity to do extraordinary things. Early on it had a bad mirror. You might remember this from the news; we fixed that. We were able to fix it because the Hubble telescope was shuttle-launched. There was a problem and we sent up a second shuttle mission to retrieve it, fix it, put it back, and come back down. It was space at its finest—actually, going up to repair something. It's like orbit is your garage. It just felt good realizing that we could do that.

It's a space-based optical observatory. We put it out in space, lifted above the atmosphere of the Earth. We've had optical telescopes for 400 years, ever since Galileo made his telescope and looked up with it. He didn't invent it, but he was the first to look up. That was 400 years ago. What could Hubble possibly do for us that we hadn't already done here on Earth? What it did for us is it enabled us to see the universe without the turbulent effects of Earth's atmosphere.

As light comes from the distant depths of space, it travels in a perfectly straight line. You get a sharp image. But, when it hits the atmosphere, variations in temperature and density force that light path to become refracted, dispersed, smudged, and smeared. By the time you take a photograph of whatever that cosmic thing was, it's a blur. It's not a total blur but small things are a total blur. Big things are okay, and you can allow a little blurring because the features are much larger than the extent to which the picture is blurred. Suppose, however, that you want to look at small features; you're hosed. You know it's there; you just can't describe it.

What Hubble enables us to do is to see the universe above the atmosphere where you don't have this blurring problem. Hubble has sharp vision, high resolution. I can analogize this with a story that was described to me. I happen to have 20/20 vision, so I found this story fascinating for someone who has uncorrected vision. The person said she got glasses very late in life. I said, "What was the effect when you first put on a set of glasses?" She said she knew, intellectually, that lawns had blades of grass but never actually saw blades of grass. It just looked like a carpet. When she put on glasses, she could see the blades of grass, collectively making a lawn. If you're denied the precision of that vision, you're severely crippled in your capacity to theorize about the cosmos. Hubble gives us that vision.

The Hubble "Deep Field" is the name of a single photograph taken by that telescope, which for me is more profound than any picture from the *Apollo* era, more so than "Spaceship Earth," "Earth Rise," even more so than the beautiful and richly detailed images of the planets in our solar system. Let me tell you what the Hubble "Deep Field" is.

First of all, if you want observing time on the Hubble telescope, what you have to do is apply. You write up a proposal, and you apply. About anywhere between two and ten times more requests for time are submitted than are allocated. It's oversubscribed. It's an oversubscribed telescope, as are most good telescopes. What you say is, "I want to look at this object because I think it's interesting for these reasons," and you go on and on and on and on. A committee reviews it and may agree that it's a good object and give you time. Suppose you presented a proposal and said, "I want to pick a part of the sky completely randomly, and I want to blow 10 days of exposures on that one spot on the sky and just see what comes up." You would be laughed out of the room because there are so many other worthy proposals, people wanting to discover the nature of quasars or black holes or the spectra of gas clouds in the universe.

What this took was something called "director's discretionary time." The director of every observatory, including the director of the Hubble telescope, has time that he or she can allocate on a whim. It doesn't have to go through peer review; they can just assign it according to their discretion. Maybe someone—if it's ground-based—was clouded out on one of their observing runs, and they have a special interest in that project; they'll give them some extra time. Maybe it's a student who needed the time for their thesis. That's typically how the discretionary time gets allocated.

In this particular case, the director said, "I'm going to take 10 days, consecutively, of the observing time of this telescope, and I'm going to select a completely random but especially boring part of the night sky." How boring is it? I don't want any bright stars there. I want it to be away from the plane of the Milky Way galaxy, because the plane is very thick with dust and gas and things that would obscure the view. I want to pick a part of the sky where Earth doesn't get in the way, because the Hubble telescope is orbiting very close to the surface of the Earth. As Hubble telescope orbits Earth, there are actually some directions in which it can look where, no matter where it is around the Earth, it can always look into that direction.

A spot was selected that was continuously visible and avoided clusters of galaxies that we already knew about, because we wouldn't learn anything. It was selected as the most uninteresting part of the sky possible. It was a little swatch of sky near the Big Dipper in the Northern Cap. At the North Pole of the Earth—on any given night the Big Dipper is tracing circles around the North Star— a continuously visible zone.

The Hubble telescope found that one spot, took 342 consecutive exposures of the identical scene. All those exposures were taken and added together to make one image, a very high-quality image. The amount of time allocated to this photo was larger than any other observing project ever granted by the Hubble telescope.

Do you know how much of the sky this image represents? The field of view of the Hubble is not that large. This particular picture was one one-hundredth (1/100) the area of the full Moon on the sky, one one-hundredth the area. To give you an idea of that, take a penny and hold it at arm's length and ask yourself how big is the size of Abe Lincoln's eyeball. That is the size of the patch on the sky represented by the Hubble "Deep Field."

Here is the image of the Hubble "Deep Field," one of the most "boring" parts of the night sky. What does it reveal: thousands of galaxies. Practically every splotch of light in that imagine is an entire galaxy not much unlike our own Milky Way, bearing hundreds of billions of stars each. What's extraordinary about this image is, given the precision of Hubble's imagery, you can get close to the picture and look in on the galaxy. I've seen pictures sort of like this before, taken by ground-based telescopes, but they are just all smudges. If you get up closer to a smudge, it's still just closer to a smudge.

Here, as I look real close, I can see structure. I see spiral arms, I see regions of extra star formation; I see different colors. Some galaxies are larger; they're likely closer. Other galaxies are smaller; they're likely farther away. Where does it get the colors? Why are some galaxies blue? Some stars are blue. Those galaxies are highly represented in that species of star. How about the red galaxies: some stars are red. Some stars are white. We have red, white and blue and everything in between. Remember this is the size of Abe Lincoln's eyeball on a penny held at arm's distance, projected on the background sky.

This is not simply a picture of something flat; it is a beam through the universe, looking at nearby things as well as objects that are the most distant, all compiled into one image.

The farthest known galaxy in all of the catalogs is in this picture. It is a red dot. How did it turn red? It's the expanding universe effect. We're in an expanding universe, and the wavelength of the light in the galaxies in an expanding universe gets stretched. If light is stretched, its energy gets lower and lower, and red light has less energy than blue light. What starts out as a blue galaxy very far away in an expanding universe, turns from blue to red as it quickly recedes. The farthest galaxy in this picture is actually red. If we zoom in on a little piece of the Hubble "Deep Field," and see the little red thing in the center, that galaxy. We see it not as it is, but as it once was nearly 13 billion years ago.

That galaxy is one of the earliest galaxies known. Its light has been traveling for 13 billion years to reach us, and it is a signpost for what the universe was like in the distant past. Once again you can see every other smudge there is a galaxy. Just to the right of our red smudge, you see structure in another spiral galaxy, very distant.

We can extrapolate our view into the Hubble "Deep Field" and say, "Suppose every patch of the sky looked like this, then how many galaxies are there?" There are about 50 billion galaxies. If you keep adding them up in a statistical exercise, several thousand here and several thousand there, you get about 50 billion galaxies. Let me reiterate that we also have ground-based telescopes, some much larger than the Hubble, like the Keck telescope in Hawaii, which is 10 meters in diameter compared to the Hubble telescope, which is about 2.5 meters in diameter. It is much bigger. It has much greater collecting power. It can detect everything that Hubble detects, but it's not seeing the detail. Without that detail the images taken by ground-based telescopes do not operate on my imagination. They influence it only barely. If you just show me a picture of smudges and tell me they're galaxies, it's one thing. But if I can see into their souls; if I can see what kind of galaxies they are—what color they are, what shape they take—and know that each one of those galaxies, based on ours as a representative case, contains 100 billion stars and that every point of light, except for one or two, is one of these galaxies; I have to take pause. I get whoopy thinking about that, the sheer scale, size and magnitude of the cosmos just panning across the image. As we pan, we see different galaxies go by and I wonder, "Are they looking at us?" Is there life on those galaxies looking back at us asking the same question? Once again by analogy to the planets, going from abstract orbs that are out there to becoming real worlds, the Hubble "Deep Field" for me, turned galaxies from abstract things out there in the cosmos into worlds, worlds unnumbered.

The Hubble telescope has, in a way, sanctified this area of the sky—the Hubble "Deep Field". What I mean by that is that people now take other kinds of telescopes and train them on that same spot in the sky to see if they can add information to it. The Hubble telescope is only an optical telescope. It's looking at red, orange, yellow, green, blue, indigo, and violet light and a little bit of ultraviolet. Let's take other kinds of telescopes and point to that same area to see if there's anything different to combine the information. The race was on to find out how many extra observations we could add to it—to create a total picture.

The Chandra x-ray telescope focused on this same area. As Hubble was a famous astrophysicist of the past, so, too, was Subrahmanyan Chandrasekhar. He was a famous astrophysicist who died recently as a Nobel Laureate. There's a telescope named in his honor, a telescope as grand as Hubble, but specializing not in visible light, but in x-rays. Notice the scene looks very different. That's good. That tells us that x-rays are looking at something different, in fact, black holes. We learned in a previous lecture that the accretion disk that surround black holes and neutron stars can give off x-rays. What we're doing now is tracking the presence of black holes, and there's no doubt that every one of these dots of light in this image is a super massive black hole in the center of a galaxy, dining happily on gas clouds that have come too close. It tells us something different about that spot on the sky.

We've got a lot of unanswered questions. Do we have the right to extrapolate to the whole cosmos what happened in this one little patch of the sky? Maybe there's something bad or non-representative about the sightlines through the Big Dipper. So we took another spot, and we did that in the diametrically opposite direction, the Hubble Deep Field South. We don't expect it to look exactly the same, because it's a different direction on the sky, but it looks statistically the same. There are about the same number of galaxies, about the same distribution of colors, about the same array of shapes and sizes. The analysis bears that forward.

We've now looked at two spots; do we now have the right to extrapolate to the whole sky? I'm extrapolating to the whole sky. I have no reason to believe that this side of the universe is going to look fundamentally different from that side of the universe. We've got it in two directions. I think it would be a waste of time to go to every single postage stamp, plus it would take a billion years to do that, given how small this region is. Let's assume that.

Questions still remain. This is such a beautiful to the edge of the cosmos, I'd ask, "Are the laws of physics the same everywhere, through all of time, in every nook and cranny?" Are there laws of physics yet to be discovered that will grant me greater insight into what is going on in this image? Just as our first views of the Moon made us want to go there to find out more about the Moon, we now know more about Earth from having been in space—to look back to Earth than did we being on Earth, looking up. Now I'm looking out, to the edge of the cosmos, and I want to know more about it. Occasionally I worry if there are enough laws of physics known to come to the understanding that I seek.

Do the planets around stars among those unnumbered galaxies have life? If they have life, is that life intelligent as we define intelligence? As I asked earlier, are they looking through their telescopes at us? Is there a Hubble "Deep Field" for their planet, and are we a galaxy sitting on their photograph? Are they asking the same questions we are, or are they just looking for shelter, food, and sex, as most life forms on Earth do most of the time? I don't know.

I do know that this Hubble "Deep Field" for me is a window, and I think of reaching out through this window. Someone has actually turned it into a window spread out in three directions. A colleague of mine, Ken Lanzetta, took the Hubble "Deep Field," and took all the galaxies, which are spread through time and distance, and created a fly-through of the Hubble "Deep Field, adding the dimension of distance to it. That makes it that much more compelling to me as an intellectual journey, as a place to gaze upon and wonder, especially since that gaze is a gaze that takes you to the edge of the cosmos.

When you go out under the night sky and look up, what are you doing? You're stargazing. When I look through this window that is the Hubble "Deep Field," I galaxy gaze. I galaxy gaze through time. In so doing, I revel in the diversity of colors and forms and shapes and structural detail exhibited as the objects of that photo. When I do this, the boundary between my knowledge of the cosmos and my ignorance calls to me. When I reach for the edge of the universe through the Hubble "Deep Field," I do so knowing that, along some paths of cosmic discovery, there are times when—at least for now—one must be content to love the questions themselves.

# Lecture Eight
# In Defense of the Big Bang

**Scope:**

This lecture is presented in defense of the Big Bang theory, which is often misunderstood and, sometimes, even discounted. Throughout time, people have asked questions about the origins of the universe and, for most of time, the answers to those questions were provided by mythology. The $20^{th}$ century was the first period in history in which we were able to use the methods and tools of science to answer our questions. We now know without doubt how the universe began, how it evolved, and how it will end.

## Outline

I.   Why do we believe in the Big Bang theory? Why do we believe that 13 billion years ago, all energy, matter, space, and time in the universe were packed into a primeval fireball, smaller than an infinitesimal fraction of the size of the point of a pin?

   **A.** Regardless of what you may have read or heard, the Big Bang is supported by an overwhelming body of evidence.

   **B.** In the 1700s and 1800s, theories were put forth in the world of physics that were tested and ultimately came to be called *laws*. Examples include Newton's laws of gravity, motion, and optics.

   **C.** In contrast, the early $20^{th}$ century saw Einstein's theory of relativity, quantum theory, and the theory of quantum chromodynamics. The terminology has changed from *law* to *theory*.

   **D.** Modern theories are just as thoroughly tested and just as successful as the ideas that were previously known as laws. In the $20^{th}$ century, however, we learned that the truths we determine about the universe may be only a subset of a larger truth.

      **1.** For example, Newton's laws of gravity and motion described the everyday environment at everyday speeds.

      **2.** In the $20^{th}$ century, physicists began to work with the concept of the speed of light, and new theories were needed to account for phenomena that occurred in this new domain.

3. Einstein's relativity describes high-speed motion. The theory of general relativity didn't force us to discard Newton, but it encloses the phenomena of the universe that Newton described and encompasses other phenomena.
4. Hence, the Big Bang is termed a theory in deference to the idea that it may someday be enclosed in a larger picture.

E. We must also note that up until the $20^{th}$ century, we examined phenomena in terms of whether they were consistent with what we would expect given the application of our five senses. For example, if we see something vibrate, we know that it will make a sound.
1. Physics in the $20^{th}$ century started using new technology, such as particle accelerators and huge telescopes, to explore domains of matter that our senses had never encountered.
2. When we start learning how the cosmos behaves at the tiny level of the particle or the vast level of the universe, our new knowledge falls outside our senses. We can no longer use the criteria of our senses to judge the implications and meanings of phenomena.
3. Theories in these realms must make mathematical sense, however, if not intuitive sense. The mathematics of a theory serves as a logical image of the idea.

II. A number of experimental pillars support the truth of the Big Bang theory.
A. The Big Bang makes three assumptions, out of which all its predictions follow. These assumptions are:
1. The universe is expanding.
2. The universe is cooling.
3. The universe had a beginning.
B. Edwin Hubble observed the truth of the first assumption in 1929 when he noticed that almost all the other galaxies in the universe were moving away from the Milky Way at high speeds. Hubble also noted that galaxies that are twice as far away from us are receding twice as fast. That behavior is a signature of an explosion.

C. Earlier, in the theory of general relativity, Einstein noted that space can be thought of as a fabric that distorts in the presence of matter. The distortion is what we call gravity.

1. Under general relativity, the expansion of the universe is not galaxies moving through a preexisting space. Instead, the galaxies are embedded in space, and what is expanding is the fabric of space, something like a three-dimensional sheet of rubber.

2. Sir Arthur Eddington, a well-known British astrophysicist, tested the theory of general relativity just a few years after it was proposed. Eddington went on an eclipse expedition, during which he measured the exact positions of stars near the edge of the Sun. On a similar expedition six months later, he took the same measurements and found that the stars were in different locations.

3. During the eclipse, light came past the limb of the Sun, and its trajectory was altered because of the curved fabric of space and time, just as relativity had predicted.

4. The results of this experiment were consistent with the theory of relativity. Individual objects curve space-time. The sum of all the objects in the cosmos curves the cosmos. Space, then, has a structure that can expand or warp.

D. How do we know that distant galaxies are receding at high velocities?

1. The speed of galaxies can be measured using the Doppler effect as it relates to light. Receding objects shift to the red part of the spectrum. The farther away an object is, the greater its shift to red.

2. General relativity predicts a phenomenon called *time dilation* for events that take place at great distances. If I'm moving away from you and counting off seconds, the farther away I move, the longer it will take my count to reach you. From your capacity to measure, the duration of a second has increased.

3. Time dilation can be measured in the universe using supernovae, which are high mass stars experiencing explosive deaths. We know how long a supernova takes to reach the height of its luminosity and how long it

takes to recede, a period known as its *light curve*. The light curve seems to be longer for supernovae in distant galaxies, but if we calculate how much longer it is, we find that the result matches exactly with how much time dilation general relativity predicts for that distance.

**E.** In the 1930s, George Lemaître, a Jesuit priest and cosmologist, was the first to connect the discoveries of Hubble and the theory of relativity. He proposed that if the universe is now expanding, in the past, it must have been smaller, and if the universe was once smaller, perhaps it had a beginning. Lemaître also proposed that when the universe was more compressed, it was hotter.

    **1.** At one point, the universe was 3000 degrees K. At that temperature, atoms are ionized, which means that electrons are detached from their host nuclei. If light tries to move through this medium, it is batted around. Light does not travel in straight lines at this temperature; it exists in a kind of fog.

    **2.** As the universe cools, electrons combine back to the atoms, and light freely flows through.

    **3.** The universe is 1000 times bigger now than it once was and is one one-thousandth (1/1000) the temperature now that it was, which equates to three degrees K.

**F.** George Gamow predicted the current temperature of the universe in 1948, a time when we had learned enough about particle physics to know how particles behave under certain conditions.

    **1.** At this temperature, objects emit primarily microwaves, not visible light. Gamow said that the signature of the Big Bang would be found in the microwave part of the spectrum.

    **2.** In 1967, two physicists at Bell Labs, Penzias and Wilson, tried to measure existing signals of microwaves that might interfere with the ability to use microwaves for communication. They found a microwave background signal, the *cosmic microwave background* that existed everywhere and couldn't be blocked.

3. Other physicists saw this signal as the visible remnant of the Big Bang, now red-shifted to microwaves.
4. How do we know that the microwaves come from the edge of the universe? The energy of microwave light is bumped up as it moves through high-temperature gas clouds in clusters of galaxies. If the cosmic microwave background is truly a background from the edge of the universe, it should have a warmer signature in these clusters, and it does. This finding proves that the microwave background comes from beyond the matter that we can see.

G. We can also look at cyanogen molecules (CN) to confirm the temperature of distant galaxies, which are also distant in time.
1. CN becomes excited in a bath of microwaves; the higher the temperature of the microwaves, the higher the excitation level of CN.
2. This excitation level in distant galaxies correlates exactly with the predicted (higher) temperature of the universe at the time of the distant galaxy.

H. One final piece of evidence is found in the levels of certain elements in the universe.
1. We know, for example, that the universe was born with 10 percent helium. Therefore, every part of the universe should contain no less than 10 percent helium, and every part does have at least that amount.
2. In the same way, lithium and beryllium were present in trace amounts at the beginning of the universe. No part of the universe should have any more of these elements than those trace amounts, and no part does.

III. Astrophysics and particle physics have joined to give us the truth of the Big Bang theory.
A. The theory still has a few problems.
1. For example, 90 percent of gravity comes from an unknown substance called *dark matter*. Similarly, the future expansion of the cosmos is driven by an unexplained phenomenon known as *dark energy*. We can't yet explain these factors in the structure of the universe.

**2.** These mysteries, however, do not cause us to discard the Big Bang theory. Perhaps a new theory will be posited that will encompass the Big Bang, just as Einstein's relativity enclosed Newton's laws of gravity.

**B.** As far as we now know, the Big Bang is completely consistent with all the data, and nothing else has ever come as close to describing how the universe is structured.

**Suggested Reading:**

Guth, Alan H. *The Inflationary Universe*. Reading, MA: Addison-Wesley, 1998.

Smoot, George, and Keay Davidson. *Wrinkles in Time*. New York: William Morrow & Co., 1993.

**Questions to Consider:**

1. What exactly is this famous cosmic microwave background? What makes it cosmic? What makes it microwave? And what makes it background?

2. What role does Einstein's general theory of relativity play in the Big Bang?

# Lecture Eight—Transcript
## In Defense of the Big Bang

Welcome back to My Favorite Universe. We're going to spend the entire lecture in defense of a very famous theory known at the Big Bang. In fact, that's the title, "In Defense of the Big Bang." I take this time to do it because it's a theory that's so misunderstood by the public. People hear the word "theory" attached to it and they say, "Oh, it's just a theory, so I don't need to believe it." Or, there's some discounting of the intellectual achievement that it represents. I am here to set the record straight.

Before I begin, let me remind you that if you look at writings throughout cultures and throughout time. People have always wondered how things began. Questions of origins seem to be perhaps even genetically encoded in our capacity to ask questions. We, living in modern times, are no exception to this. Even today, when I'm on an airplane and people find out I'm an astrophysicist, they ask me— it's not the first question they ask, the first is, is there life elsewhere in the cosmos—the second is, "Tell me about the Big Bang." So it's there—people want to know.

The 20$^{th}$ century was the very first time in history when we were able to bring the methods and tools of science to answer that question. All previous times, and all previous cultures, had no greater tools than simply the mythology of their imagination. We do live in special times, because we're not relegated to just having to imagine how things began. We can look out with out largest telescopes, probe exotic states of matter with our particle accelerators and come to terms with how the universe began, how it evolved and how it will end.

Let's explore the evidence for the Big Bang. Let me begin by saying, "What could possibly force a rational astrophysicist to believe that, all the matter, energy, space and time of the universe began 13 billion years ago—packed into a primeval fireball smaller than some infinitesimal fraction of the size of the point of a pin—and it's been expanding ever since?" What could possibly bring someone to believe in that? The answer is simple. Regardless of what you may have read or heard, the Big Bang is supported by an overwhelming body of evidence.

There is more evidence in support of the Big Bang than any idea that has ever preceded it or any idea that has come since. It's all about the evidence; it's not about belief systems. I've had people come up to me and say, "Do you believe in the Big Bang?" I don't even understand the question. The question properly worded would be, "Of all the evidence that's out there, what idea is best supported by that evidence?" It's the Big Bang.

If you look back in the 1800s and the 1700s, there were theories put forth in the world of physics that, while some were radical ideas, were finally tested and confirmed, and then they got called "laws." There's Newton's law of gravity, Newton's laws of optics, laws of motion, laws of thermodynamics. It was the era of laws. When you enter the $20^{th}$ century—very early in the $20^{th}$ century we get to Einstein's theory of relativity. In the 1920s we have quantum theory—quantum chromodynamics—all these theories. We no longer use the term "law," and that's for a couple of reasons. First, these modern theories are as tested and as successful as what were previously known as "laws." We could call them "laws" if we wanted and be consistent with the naming schemes of our generations past.

What happened in the $20^{th}$ century was that we came to learn that, whatever it is we determine to be true about the universe might only be a subset of a larger truth. We learned that with Newton's laws of gravity. Newton's laws of gravity and laws of motion described the environment in which Newton was comfortable, where he had never seen anything move faster than a fast running horse. His laws of motion surely accommodated anything that came up in his day. Enter the $20^{th}$ century and we start thinking about the speed of light, high-speed motion, and all kinds of other phenomenon that take place. We needed new theories to account for these new domains in which we were testing how the universe works, and Einstein's theory of relativity does just that; it's a theory of motion and of gravity. It encloses the piece of the universe described by Newton. It doesn't discard Newton; it enhances the sphere over which the laws apply.

In modern times when we say "theory"—if it's a well-tested theory—had that been the same theory back in the 1800s, they'd be calling it a law, as they'd be calling the Big Bang, the "Big Bang Law." I will remain humble in the presence of theories yet to be put forth, recognizing that perhaps one day the Big Bang will be enclosed in a bigger picture, a deeper understanding of how the universe works. The day that happens we're not throwing away the Big Bang. We will preserve it as a piece of what works under much broader circumstances and under much wider conditions.

There's another piece of information we need before we start analyzing the evidence, and that is that we're born and raised with our five senses: sight, hearing, touch, taste, and smell. As you grow up, you receive things that happen through these senses, and you come to expect things to follow certain rules and regulations. If you drop a cup, it falls. If something vibrates quickly, it makes a sound. These things make sense. The definition of "make sense" is, they are consistent with what we expect to happen given the application of the five senses with which we are born.

Enter the $20^{th}$ century where science, physics in particular, no longer takes place on the tabletop. We start getting into particle accelerators, exploring domains of matter that our senses have never seen before, whipping out the largest telescopes in the land, looking to the outer reaches of the cosmos, experiencing a scale of the universe unlike anything we have ever seen in our lives. We start deducing the nature of the cosmos, learning how the cosmos behaves, and it's falling outside of our senses. No longer can we invoke the does-it-make-sense criterion for theories or phenomenon that we see take place in the cosmos. It is long past our capacity to judge whether or not something makes sense.

Throughout all this, we've always required that whatever is your theory; it ought to make mathematical sense. The math is a logical image of the idea, so you can manipulate the idea—ensuring that you're not going to make a mistake in so doing. It's always got to make mathematical sense; but it doesn't have to make intuitive sense, because your intuition was carved in the wrong place. We also demand that, if you come up with a new theory, it has to agree with the evidence and predict things that maybe you hadn't known before. Otherwise, it's just sort of an explanation that comes after the fact, putting pieces together without giving you new insight into how the world works.

For the Big Bang in particular, a variety of experimental pillars support its status as the most successful theory every put forth. I call them pillars because the thing is resting on these pillars. The Big Bang makes three assumptions from which all the predictions that it makes follow. One assumption—although it was known at the time the theory was put forth—was that the universe is expanding. The second assumption: Yesterday the universe was hotter than it is today. If you turn the clock back and you somehow found a way to measure the temperature of the universe, you'd get higher temperatures than now. The third assumption asserts that the universe had a beginning. That's all it needs, those three assumptions. By the way, if it explained everything we knew and it had a thousand assumptions, that's not a very good theory. A theory should have the fewest number of assumptions possible. That's what gives you confidence in its power.

Let's take these one by one. In 1929 Edwin Hubble of Hubble telescope fame, discovered that all the galaxies—except for a few that are sitting as our neighbors—he saw in the nighttime sky had high speeds and all those speeds were moving away from the Milky Way galaxy. All the galaxies, except for a handful close to us, were expanding away from us. That's something he noticed. He also noticed that galaxies that were twice as far away were receding from us twice as fast—and three times farther away, three times as fast. It wasn't just sort of a uniform expansion; there was this stuff far away that was moving faster. That is a signature of an explosion. If you studied the debris tossed in any explosion, what you would find is that some debris landed twice as far away as the other debris, yet it all settled at the same time. It meant that the farther debris had twice the speed as the debris that was only half as far. It is the signature of an explosion.

Thirteen years before this, Albert Einstein put forth a new theory of gravity, *general relativity*. General relativity generalized the assumptions of what went into the special theory of relativity. The special theory had constant motion and there was acceleration but there was no discussion of gravity within it. Gravity causes accelerations. It was an important case of a broader story called the "general theory of relativity." In the general theory of relativity he noted that space, the space through which we walk, actually could be thought of, as a fabric that distorts in the presence of matter, and the distortion is what we call "gravity."

This expansion of the universe was not a galaxy moving through a preexisting space. Under the tenets of general relativity, the space and the galaxies within in it are embedded together, and it's the fabric of space that is expanding—like some kind of three-dimensional rubber sheet. They say that doesn't make sense, but it doesn't have to make sense, like I said. We're talking about modern physics here and domains that don't happen in your everyday life. It didn't happen in a sandbox when you were growing up.

The general theory of relativity was proposed in 1916, and in 1919—three years after it was proposed—it was already tested. Sir Arthur Eddington, a well-known astrophysicist in England, went on an eclipse expedition, waited for the Sun to be blocked by the Moon, saw stars come out in its vicinity and measured the exact positions of those stars, right near the edge of the Sun. You can't see them when it's not an eclipse, so obviously he needed this condition to make the measurement. He waited six months, then took the same measurement and noticed that the stars were in different places. During the eclipse the light came past the limb of the Sun, and its trajectory was altered because of the curved fabric of space and time in the vicinity of the Sun, just as predicted by Einstein's relativity.

Relativity, as exotic as it was, was turning out to be correct. While individual objects curve space-time, the whole universe—the sum of all the objects in the cosmos—curves the entire cosmos. Space is something that has structure that can expand or curve or warp, and galaxies are embedded within that space. It is consistent with relativity.

How do we really know that distant galaxies have high velocity? We measure using something akin to the Doppler effect—like the police radar gun. They aim the radar gun at you and they measure your speed of approach or speed of recession. In the galaxy, if you perform a similar kind of experiment using light of any kind, you notice that features in the spectrum are shifted toward the red part of the visible spectrum. There is red, orange, yellow, green, blue, and violet. If there is any feature in that spectrum that belongs attached to a particular color, if an object is receding, it shifts toward the red part of the spectrum. We call it a "red shift." The farther away the object, the greater will be the red shift measurement.

For objects that are receding fast, general relativity predicts that there is something called a *time dilation* for events that take place at these great distances. Let's say I'm moving away from you, and I'm counting out seconds: one, two, three, four. If I'm receding from you while counting out seconds, with every second I lay down I am farther away from you than I once was, and that signal doesn't get to you until later, so the duration of a second is actually increased according to your capacity to measure. It's called "time dilation." It's a prediction of general relativity.

Sure enough, we have a way to measure this without actually going out into space. There are objects called "supernovae." Here is an example of one in a nearby galaxy. It is that bright star in the bottom left corner of the image. It looks like a star that's brand new on the scene, but, in fact, it's a dead star. It's the explosive death throes of a high-mass star, thrusting its chemically enriched guts into the galaxy. Sometimes they are so bright they rival the luminosity of the 100 billion stars in the galaxy in which they reside. We understand these beasts. We know how quickly they gain in luminosity. We know how slowly they decline and how long that takes.

Let's look farther out in the cosmos. If we look farther out we find much more distant galaxies, now much smaller, they're just smudges. They've got supernovae, too. If you look at each one of these images you see a bright star, rivaling the brightest of the rest of the system. These supernovae are far enough away so that this effect of general relativity can be measured. We look at their rise in brightness and their drop off, and we say, "You know something? All the nearby supernovae rise up and drop down in this much time, but these take longer to execute their course of brightening and dimming." Maybe it's a different species of star in the early universe; it just has a different property. Maybe. Let's figure out how much longer the period we call *light curve* lasts. Calculate that. It's easy to just measure it up. When we compare it to how much time dilation general relativity predicts for a galaxy at that distance, it's bang on. It's exactly what we measure with the stretching of the light curve of the supernova. We measure the time dilation. That's real, no doubt about it. Let's put this together. We've got an expanding universe. We've got galaxies embedding in space. We've got a theory to account for it, the general theory of relativity.

There was a brilliant man around at the turn of the century, but the work for which we remember him was conducted in the 1930s. He was George Lemaître, a Jesuit priest and cosmologist, impressive things to have on your business card. He was the first to connect the new theory of relativity, the new discoveries of Edwin Hubble, and the fact that you've got galaxies sort of embedded in this expanding system. He was the first one to think, if we're expanding, maybe in the past we were smaller; and if we were smaller, maybe we had a beginning. Just maybe the universe had a beginning. Maybe, based on his knowledge of physics, the universe was hotter at a time when it was much more compressed.

If you've ever pumped up air into a bicycle tire, if you feel the valve, it's hot after you've done your pumping, compressing air into a smaller volume. If you let the air out with your thumb—if you don't have a bicycle, just try it on a car tire; it works, too—when the air comes by it will feel cooler. Rapidly expanding air drops a little bit in temperature. The expanding universe is dropping in temperature. It operates under similar physical principles.

If you turn the clock back, let's look at the universe. It was hotter, hotter, hotter, hotter, and hotter. There was a point where it was about 3000 degrees. At 3000 degrees you're aglow; and if you're at 3000 degrees, atoms are ionized, which means the electrons are detached from their host nuclei and they're sort of swimming in this soup. When light tries to move through the medium, it gets batted back and forth, so that light does not travel in straight lines through the universe. It's kind of a fog of light, actually. It was 3000 degrees at a particular time in the history of the universe. At all times before that the universe was opaque; but at that point, as we cool through those temperatures, electrons combine back to the atoms and light freely flows through.

It's been a long time since that happened. The universe is 1000 times bigger. How much cooler should we be? It is proportioned, so we're one one-thousandth (1/1000) the temperature that it was back then—now being 1000 times larger. It was 3000 degrees then. Right now the temperature of the universe is three degrees Kelvin (K). Kelvin is the absolute temperature scale. If you're zero on the Kelvin scale, there's nothing colder. At zero on Celsius, you've got colder temperatures; zero on Fahrenheit is really cold, but you can get colder. At zero on Kelvin, that's all she wrote—three degrees Kelvin in the universe, that's the measured temperature today.

This bit of arithmetic was actually first done by George Gamow after the Second World War, in 1948. We knew enough about particle physics having just finished the war effort and the Manhattan Project, building the bomb and getting inside the atom. We knew enough about the atom and about particles and particle physics to know that particles behave in particular ways under particular circumstances. The electron would jump out of the atom and back into the atom and light would be released.

He predicted what the temperature should be today. He said it would be about five degrees K, plus or minus. He got the wrong answer, but his idea was right. At five degrees are you emitting visible light? No. Your photons are primarily microwaves. He said maybe someone will look around for this signature of the Big Bang, and it will be found in the microwave part of the spectrum.

In 1967, 19 years after he made this prediction, two physicists at Bell Labs, Arno Penzias and Robert Wilson—Bell Labs is the communication center of the United States and it is where much of the modern technology that we use to communicate with each other was invented, including, for example, the transistor. Penzias and Wilson were trying to measure what signals of microwaves exist in the sky that might interfere with the ability to use microwaves to communicate. They brought out their antenna to measure and removed everything they could possibly find, but there was still a signal left over that they couldn't remove. It was everywhere, no matter where they looked. It didn't focus in one direction and not another; it was everywhere.

They published it. They said, "Excess microwave signal everywhere we look." They were not cosmologists. They were sort of engineering physicists. It took other physicists to look at that signature and say, "You've discovered the Big Bang." The visible remnant now red-shifted to microwaves. It was so red-shifted it was out of the visible spectrum, past infrared, all the way into microwaves. It shifted into that part of the spectrum, and is what George Gamow predicted. You have just seen evidence of the birth of the cosmos. They won a Nobel Prize for that discovery, even though they didn't know what they had found—which is kind of cute. They slipped into the Nobel Prize, but somebody had to discover it.

How do we know that microwaves actually come from the edge of the universe? Maybe they're scattered or are somehow a property of space. If you look between here and there you'll see a little less than if you look between there and there, because you're not looking through the whole distance. There is a way to test for that. Let's look at a cluster of galaxies. This image shows a chevron because it's the shape of the detector for the Hubble space telescope, and it is an image of a cluster of galaxies. Every smudge in the image is a galaxy. We know that clusters of galaxies have gas within them that is extremely hot, so hot that it's radiating x-rays. When microwave light, coming from the edge of the universe, comes upon one of these clusters, the high-temperature of the clusters actually gives a kick and bumps up the energy just a little bit. If the cosmic microwave background is truly in the background, every place where there is a cluster, the microwave signature will be a little warmer for having passed through it.

The x-ray feature in the picture from the Chandra x-ray telescope shows a glowing gas and also a signature of black holes, which appear as two bright spots in the centers of galaxies; but all the rest is gas. Sure enough, everywhere we look, every sightline in the sky that passes through one of these clusters that has a hot, intra-cluster x-ray-emitting gas, the cosmic microwave background is pumped up a little bit in its temperature. That tells us that this stuff is coming from beyond every piece of matter that we see. That's an important distinction.

Is there another way to know that it was hotter in the past than today? That is one of the predictions. If you look far enough back in time at galaxies out there—and just by looking up you are looking back in time because it took light time to reach you—you can find a galaxy anywhere you want. Go back in time a billion years, two billion years; whatever time you go back, the universe was hotter then. There is a molecule, the cyanogens (CN) molecule, that gets excited when taking baths in microwaves, and the higher the temperature bath of microwaves, the higher is the excitation level of that molecule. We can measure that here and find the temperature of the bath is about three degrees. If you find that cyanogen molecule at a distant galaxy and measure it, it's 10 degrees. Is that consistent? Figure out how much size the universe changed between now and then, correlate it with temperature, flip in the equation, bam; it's exactly the temperature of the universe at that time—a smoking gun.

What else do we know? The universe was born with a lot of hydrogen and 10 percent helium. It was born that way, so we'd expect every nook and cranny of the universe to have no less than 10 percent helium. We looked because that was predicted out of the early universe nucleosynthesis—the early universe particle physics. Regimes of laboratory physics tell us what was going on then, and it predicts 10 percent helium atoms anywhere we look. There could be more, but nowhere will have less because that was what the galaxy was born with. We look around the galaxy and no part has any less than 10 percent helium, no less. That's a smoking gun if I ever heard of one.

There are a couple of other elements, lithium and deuterium made in trace amounts. They have the opposite problem. They're not made in any great way after the Big Bang; in fact, they're easily destroyed. So, what we look for there is that you don't expect more than what the universe was born with, because natural phenomena in the universe destroy it. If you look around, you don't see any more of it than what was predicted in the Big Bang.

We have two predictors that actually work in opposite directions: the helium—no less than 10 percent, the lithium and deuterium—no more than their miniscule fractions. We don't expect any more than one in one hundred thousand hydrogen atoms to be deuterium. Deuterium is just a version of hydrogen that has an extra neutron in its nucleus.

We didn't just make this stuff up. There was an unprecedented marriage of astrophysics and particle physics, and a coherent picture has emerged by the application of those two disciplines. When you combine them, it tells us that the galaxy velocities are real. The galaxy distances are real. The expanding universe is real. Relativity is real. Quantum mechanics is real. The early universe was hot and the Big Bang is law.

All is not golden; there are some problems that worry all of us. Ninety percent of all the gravity in the universe comes from something that we don't even know what it is. We call it *dark matter*. We don't know what's causing it. Also, the future expansion of the cosmos will be driven by some phenomenon that acts opposite the sense of gravity. We call it *dark energy*. We don't know what it is. It's like an anti-gravity pressure. We don't know what that is. By the way, there's a place for it in Einstein's equations, but just because it has a place in the math doesn't mean we understand the phenomenon.

We can't use the Big Bang to tell us what happened before the first 10 million, trillion, trillion trillionth of a second. That's a long time ago, but at that moment general relativity and quantum mechanics break down. They're incompatible at times before that. We need a new theory to get us before that. The Big Bang also doesn't tell us what happened before the beginning of time itself.

Do we throw away the whole theory just because it doesn't fit? No, that would be unwise. What if we had done the same thing with Copernicus, who put the Sun back in the center of the universe—instead of the Earth—and had planets orbiting in perfect circles? He had the basic idea right; the details were wrong. The orbits were ellipses, not perfect circles. You don't throw away the whole idea if the idea has the correct thing going on in its soul. Then, you fix the details around the edges.

I look forward to the day when maybe a new theory supplants the Big Bang and the Big Bang is just part of a bigger story. There's one going around right now. If you look back early enough, this sort of quantum foam—this gurgling of the early universe in its earliest times—actually spawns different universes, an unlimited number of universes, bubbles growing out of this hyperspace. Our bubble happens to be our universe with our laws of physics. Other bubbles might have different laws. Maybe that universe is incompatible with life, so we're not there making these statements; we're here. In that case, the beginning of our universe is just part of this long-going, long-running phenomenon, making baby universes daily. There's not much evidence in support of those ideas, but they're intriguing nonetheless, and they're consistent with some major tenets of relativity.

I'll leave you with some thoughts. First, whether or not you like the Big Bang is irrelevant. It's consistent with the data, and the data overwhelmingly support it. Nothing's ever come as close to describing how the universe is structured as the Big Bang, and science does not discard successful theories; it enhances them. I leave you with a quote from Max Planck, a famous physicist from 1936. He said, "An important scientific innovation rarely makes its way by gradually winning over and converting its opponents. What does happen is that its opponents gradually die out and that the growing generation is familiarized with the idea from the beginning."

# Lecture Nine
# The Greatest Story Ever Told

**Scope:**

This lecture synthesizes the greatest discoveries of physics, astrophysics, chemistry, and biology to present a coherent story of the birth and evolution of the cosmos. Modern humans are not the first group of people to speculate about cosmic evolution, but we are the first to use the tools of science to describe the birth of the cosmos, trace its progress, and understand our place in it. This lecture brings together all the branches of science to tell the story of our existence.

## Outline

I. Before we begin, we must have a basic understanding of Einstein's famous equation, $E=mc^2$.

    A. This equation allows us to calculate the energy equivalent of the mass of a particle, such as a proton. We multiply the mass of a proton by the speed of light squared to arrive at the energy equivalent.

    B. This process also works in reverse. If you begin with a concentration of energy, you can use this equation to determine what kind of particles can be produced.

    C. The conversion of matter to energy and back again was rampant in the early universe and can be found today in particle accelerators. It no longer commonly occurs in the everyday world.

II. Twelve to fourteen billion years ago, all space, matter, and energy in the known universe was contained in a volume less than one-trillionth the size of the point of a pin, or about the size of an atom.

    A. Conditions at that time were so hot that the basic forces of nature that collectively describe the universe were unified.

        1. Unification of forces is not a new concept. In the 19th century, electricity and magnetism, which had previously been thought of as two separate forces, were revealed to be two sides of the same coin.

**2.** A theoretical formalism enabled us to see these two forces as one force manifested in different ways.

**B.** For reasons unknown, the point that contained all the universe began to expand.

    **1.** Black holes were spontaneously forming and disappearing out of the energy contained in the unified field.

    **2.** The energy density was so high that the result was not the formation of particles, but the formation of black holes.

    **3.** As we know, black holes are the curvature of space and time. The fabric of the universe reacts to the presence of black holes.

    **4.** If black holes are forming and unforming spontaneously in a small volume, the structure of space and time becomes severely curved and transforms into a spongy substance called *quantum foam*.

**III.** Quantum mechanics, which was developed in the 1920s to describe matter at its smallest scale, and Einstein's general relativity, the modern theory of gravity, do not intersect. These two theories operate in different domains.

**A.** Even though these theories seem unrelated, they must have been unified when the universe was the size of an atom.

**B.** As the universe continued to expand and cool, gravity split away from the other forces of nature. Next, the strong nuclear force and the electroweak force split from each other.

**C.** We believe that this splitting was accompanied by an enormous release of energy that had been stored in these merged fields. This energy forced a rapid expansion of the size of the universe equal to $10^{30}$.

    **1.** This process is similar to a phenomenon that is observed in chemistry when water is frozen. The temperature of liquid water placed in a freezer will drop until it reaches a certain point when the water is converting itself to ice. After the liquid water is completely ice, the temperature will begin to drop again.

    **2.** At the point where the temperature is not dropping, the latent heat of the liquid water is being released.

**D.** The rapid expansion caused by the splitting of the fields stretched the cosmos. Much of the distinct variation in density and form of the universe became softened at this time.

    **1.** That smoothing of the universe is now down to one part in 100,000.

    **2.** Imagine creating a ripple in a two-mile–wide lake. A ripple corresponding to the smoothness of the universe would be only an inch tall.

**E.** When it formed, matter coalesced in the ripples, or fluctuations, in the universe. In the large-scale structure of the cosmos, the galaxies are in those ripples.

**IV.** At this time in the evolution of the universe, the temperature was hot enough for photons, that is, particles of light, to spontaneously convert their energy into matter/antimatter pairs, such as a proton and an anti-proton.

**A.** The result was a soup of matter/antimatter and photons, an interplay of particles forming, then annihilating, then re-forming.

**B.** For reasons unknown, matter/antimatter symmetry was broken. One out of every billion photons that would normally convert itself into a pair made a single particle of matter.

**C.** As the universe continued to cool, the energy level of the photons dropped below that required to create particles. What was left, then, was photons and matter particles.

**D.** These particles are responsible for all the structures of matter that we know of in the cosmos, including stars, galaxies, and light.

**V.** The four separate forces that were unified in the early cosmos are the strong nuclear force, the electroweak force, electromagnetism, and gravity.

**A.** The strong nuclear force binds particles in the nucleus of the atom.

    **1.** Protons are positively charged and repel each other, but they exist together in nuclei. How?

    **2.** If the protons get close enough to each other, the strong force takes over and binds them together.

**B.** The electroweak force is responsible for decay of nuclei.

**C.** Electromagnetism is revealed in the bonding of atoms; this force keeps matter together.

**D.** The force of gravity has the greatest impact on events in the cosmos.

**VI.** The temperature 300,000 years after the Big Bang was 3000 degrees K.

**A.** Until the universe reached that temperature, protons and neutrons were combining to make the lightest elements on the periodic table—hydrogen, helium, and lithium. Electrons were roaming free. The temperature was not cool enough for electrons to settle into atoms.

**B.** When electrons and photons interact, light is scattered, which means that at temperatures hotter than 3000 degrees, the universe was opaque. Our telescopes today cannot see through that wall of light.

**C.** At 3000 degrees, electrons became bound to atoms, and the universe became transparent. The photons, which have continued to cool, still exist in the universe as the cosmic microwave background.

**VII.** In the first few billion years of the cosmos, 50 to 100 billion galaxies were formed, each containing up to 100 billion stars and each star undergoing thermonuclear fusion in its core.

**A.** The pressure and temperature conditions in stars of more than about 10 times the mass of the Sun are great enough to create heavy element factories in their cores. These *supernovae* create carbon, nitrogen, iron, oxygen, and so on, then explode and scatter the elements throughout the universe.

**B.** After about seven or eight billion years of this chemical enrichment of the universe, an undistinguished star was born—the Sun. The gas cloud that made the Sun had a big enough supply of heavy elements that it spawned a system of planets, tens of thousands of asteroids, and trillions of comets.

**C.** The Earth spent the next 600 million years under heavy bombardment from the debris that was left over after the formation of the solar system. The asteroids pounding Earth raised the temperature high enough to render the planet sterile. Earth had the ingredients for life, but nothing could form.

**D.** When the bombardment subsided, Earth's surface cooled, and complex molecules began to form in the chemically rich liquid of the oceans.

   **1.** Two hundred million years elapsed from the end of the period of heavy bombardment to the point of Earth's oldest fossils, dating to about 3.8 billion years ago.

   **2.** The earliest life forms were single-celled anaerobic organisms. They thrived in an atmosphere of carbon dioxide and released oxygen as a waste product.

   **3.** The release of $O_2$ enabled a population of aerobic creatures to take over Earth. The release of $O_3$ enabled the formation of the ozone, which protects us from the ultraviolet radiation of the Sun.

**E.** We owe the remarkable diversity of life on Earth to one element, carbon. Carbon is plentiful in the universe and can make more kinds of molecules than all the other elements combined.

**F.** As we've learned, though, life is fragile. An asteroid hitting Earth could change the entire ecosystem and cause the extinction of 70 percent of the species of life on the planet.

   **1.** Of course, such an event occurred 65 million years ago, causing the extinction of the dinosaurs.

   **2.** That occurrence, in turn, opened up a new ecological niche for the survival of mammals. One branch of mammals became primates, which evolved into Homo sapiens.

   **3.** That life form invented the methods and tools of science with which to deduce the origins and evolution of the universe.

**G.** This progression tells us that we are not *in* the universe but *of* it. We are born from the universe, and we have been empowered to learn about it and figure it out—and we've only just begun to do so.

**Suggested Reading:**

Anonymous. *The Bible According to Einstein*. New York: Jupiter Scientific Publishing Co., 1997.

Schulman, Eric. *A Briefer History of Time*. New York: W.H. Freeman & Co., 1999.

**Questions to Consider:**

1. What areas of profound ignorance remain in the greatest story every told?

2. What was so important about the first three or four minutes of the universe?

# Lecture Nine—Transcript
## The Greatest Story Ever Told

Welcome back to My Favorite Universe. In Lecture Nine we are going to do something special. I am going to present to you what I like to call "The Greatest Story Ever Told," and in it I have the privilege of synthesizing for you the greatest discoveries of physics, of astrophysics, chemistry and biology. To present to you a coherent story of the cosmos, from the earliest times right up to the present. We're not the first to have wondered about cosmic evolution. If we go back to Lucretius, he says, "The world has persisted many a long year having once been set going in the appropriate motions. From these, everything else follows." How prescient that quote was—given how long ago it was written—to realize that the universe is a describable place using the methods and tools of science, which allow us to determine how it was born, how it evolved and where it will go in the future.

Before I begin this story, "The Greatest Story Ever Told," I want to first describe a famous equation. I think it's the first equation we all learned in elementary school. It's $E=mc^2$. We learn that equation before we even know what it means. We know it has something to do with Einstein. It has a lot to do with the birth of the universe. Let me remind you if you once knew—and I will tell you for the first time if you've never known—that, if you have particles—take a proton, for example—you can actually ask the question, "What is the energy equivalent of the mass of that proton?" That's a perfectly legitimate question. And, do you know how much energy is contained in that proton? Plug it into that equation and find out. Plug in the mass on one side, multiply it by the speed of light squared (that's the "$c$") and out, on the other side, you find out if you converted that proton into energy, that's how much energy you're going to get.

It works in the reverse. If you have concentration of energy and you want to ask, "What kinds of particles can I make with this energy?" Go look up on the table and find out which particle has about the same energy equivalent as the energy you have, and then you can make that particle. This two-way street that exists between matter and energy as described by $E=mc^2$, is rampant in the early universe. It's rampant in particle accelerators. You just hardly ever see it at home or in the street or on the sidewalk or in the park.

Let us begin, now armed with that important piece of physics. Where else shall we begin but in the beginning—some 12 to 14 billion years ago? All the space and all the matter and all the energy of the known universe were contained in a volume less than one-trillionth the size of the point of a pin. That size is about the size of an atom. Imagine the entire cosmos compressed down to something that small. Conditions were so hot that the basic forces of nature that collectively describe the universe were unified—they were one.

The unification of forces is not a new concept. In the 1800s, electricity was known as a force, as was magnetism—until later in the 19th century it was revealed that, in fact, electricity and magnetism were two sides of the same coin. There was a theoretical formalism, which enabled us to look at those two forces as the same thing, manifested in different ways just depending on the circumstance. That was the first time two forces were unified: electricity and magnetism becoming electro-magnetism. That's an example of a unified force. We have reason to believe that all the known forces of nature were unified in these early times in the cosmos.

For reasons unknown, this pinpoint cosmos began to expand rapidly. Before the time when the universe was a rather warm million trillion, trillion degrees and a rather young one ten-million, trillion, trillion, trillionth of a second old, all of our theories break down and we have no clue what was possibly happening in the universe before that time. Black holes were spontaneously forming and disappearing out of the energy contained in that unified field. The energy density was so high that it wasn't just making particles—switching back and forth on this two-way street between matter and energy—it was making entire black holes.

Black holes, as you know from an earlier lecture as described by Einstein, are the curvature of space and time. The fabric of the universe is reacting to the presence of black holes. If you have black holes forming and un-forming spontaneously in the early universe all contained in a small volume, the structure of space and time becomes severely curved and it gurgles into a spongy, foam-like structure. The laws of quantum mechanics tell us this, and we have a term for it; we call it *quantum foam*. During that period, a fascinating thing was happening.

The subject of quantum mechanics, discovered and invented in the 1920s to describe matter on its smallest scales, and general relativity, the modern theory of gravity put forth by Einstein, describing the nature of the entire cosmos; those two theories do not intersect. They operate on completely different domains of the cosmos. One is the smallest; one is the largest. We're worried about that because you can't look at one theory and infer the existence of the other; yet, they both accurately describe the cosmos. We're worried about it because, how can two things that don't know about each other be simultaneously true in the same universe? So, we're working on that.

What we know is that, back in the early times of the cosmos, the whole universe was small. We have this kind of shotgun wedding between quantum mechanics and general relativity. They can't avoid looking like each other because the whole universe is down to the size of an atom—an atom containing the entire universe. Quantum mechanics and gravity were one; they had to be. There's nothing else they could have been but one, under those conditions.

As the universe continued to expand and cool from those times, gravity split away from the other forces of nature. As forces split from each other, they began to occupy the domains in which we have come to know that they matter. Gravity split away and started looking like a force unto itself, with no identity shared with the other forces. Quickly after that, the strong nuclear force and the electroweak force split from one another. What happened then? We believe that was accompanied by an enormous release of stored energy, energy that was stored in the merged fields. When these fields split, energy was released, so much energy that it forced a $10^{30}$ expansion on the size of the universe—a rapid expansion. We call that the *epic of inflation*.

This release of stored energy is not unlike the latent heat of freezing that you may remember from your chemistry class. It's very much like that. If you have water and you put it in a freezer and watch the temperature of the water drop, there's a point where the temperature stops dropping because it's now converting itself into ice. When it's completely ice, the temperature drops again. What was going on in there? You're trying to take heat out of the system and how come the temperature didn't continue to drop throughout the time it was becoming ice? It's because there is a latent heat in the liquid form of that water that is not there in the solid form of that water. That latent heat got released into the environment. No, making ice cubes doesn't make the universe expand, but when you're splitting forces in the early universe, you have the kind of energy that does impose itself on the structure of the cosmos.

This rapid expansion stretched the cosmos. If you stretch the universe by $10^{30}$, much of the variation in density, in structure and form is affected. Whatever might have been a heavy fluctuation in the early days becomes softened, and we've measured this. That smoothing of the universe is now down to one part in 100,000. How smooth is that? Imagine a two-mile-wide lake and you put a ripple in that lake. The ripple would be an inch tall compared to the size of the lake—an inch tall on a two-mile-wide lake. That's a fluctuation of one in 100,000.

We'll get back to those fluctuations, but it turns out to be significant, because it's in those fluctuations where ultimately matter, once it finally forms, will coalesce. When we look at the large-scale structure of the universe and ask where are all the galaxies, we're going to find those galaxies where those ripples were. The imprint of what the later structure of the cosmos will be is being shaped in this time.

Continuing onward with what now are laboratory-confirmed physics—rather than the theoretical speculation that has come before this—the universe was hot enough for photons, particles of light, to spontaneously convert their energy into matter/antimatter pairs. Light loves to do that when it has enough energy. Examples of matter/antimatter pairs are a proton and an anti-proton, a neutron and an anti-neutron. Pick your particle; it will make a matter/antimatter pair. By the way, antimatter is real. It was all over the early universe and we make it in particle accelerators. It's not just an invention of science fiction; it is real. It's just not familiar in everyday life. It's not often you have an encounter with antimatter; if you did, it would be your last encounter.

There is an interplay between the photon, which has so much energy it can start spawning particles, and the matter/antimatter pairs, which can form the photon again if they get together. We have a two-way street going back and forth. It's a soup; a matter/antimatter and photon soup, once again invoking $E=mc^2$ to make that happen.

For reasons unknown, matter/antimatter symmetry in these conversions was broken. This is profound. We don't know why that symmetry was broken, but it was broken. What that meant was one out of a billion photons that would normally convert itself into a pair, made just a matter particle without the antimatter counterpart. It's a broken symmetry. In other words, for every billion matter/antimatter particles on this two-way street, there's one matter particle that has no mate. The symmetry was broken.

As the universe continued to cool, the matter/antimatter got together. It's like a square dance. You pair up with people, and if you're the odd one out—everyone else is paired up—you're left over. These matter/antimatter pairs made photons and the photons continued to cool and dropped below the energy level to make particles. We just froze out some photons from what was formerly a two-way street, got rid of the 100 billion paired matter/antimatter particles, leaving a matter particle and a billion photons. It's these matter particles that are responsible for all the structure of matter that we know of in the cosmos: the stars, the galaxies and the like.

These forces seemed to have come out of nothing in the early universe. In modern day they are revealed as four separate forces having started as one in the early universe. There's the strong force already mentioned. That force binds particles in the nucleus of the atom. Protons have positive charges and they're sitting next to each other in atomic nuclei. How is that even possible—because like charges are supposed to repel? In fact, they do repel unless they get close enough and the strong force takes over. That's why we call it a strong force. It's stronger than their tendency to want to repel each other. The strong force keeps nuclei together.

The weak nuclear force is responsible for the decay of atomic nuclei. There's a third force, electromagnetism. As I said, back in the 1900s they were two separate forces where they are now one—one force even in our own lives. Electromagnetism is revealed in that which keeps matter together, the bonding of atoms: human flesh, wood, carpet; it doesn't matter. Electromagnetic forces bind things together in our everyday lives. The fourth force, the one that has the largest range, the greatest impact on the events of the cosmos, is gravity.

While the energy of the photon continued to drop—because the universe is expanding and cooling—the photon can no longer make particles. It's not possible any more. Gone is the era where particles were manufactured out of the energy of the universe. It stops. It will still happen in very restricted places, like the centers of stars.

The universe continues to cool 300,000 years after the Big Bang. We went from those very high temperatures to about 2000–3000 degrees K. now. Up until then, protons and neutrons had combined to make the lightest elements on the periodic table. They made hydrogen, helium and even lithium. The electrons were free to roam. It was not cool enough yet for an electron to settle down into the atom. We had nuclei flying around and electrons flying around.

Light likes electrons. If an electron sees a photon of light, it will bat it back and forth. They interact with each other, so they scatter the light. Light did not have a free trip through the universe. It was like a random walk, like a pinball moving through a pinball machine. Before 300,000 years, the universe was opaque. Our telescopes today cannot see through that wall of light. It's like the frosted glass in a bathroom window. At 300,000 years the temperature dropped to a couple of thousand degrees and that was cool enough for the electrons to be bound to the atoms, and the universe became transparent. All of the photons left over from that period were still in the universe, but as the universe expanded, they continued to cool. At what then was a temperature represented as 2000 or 3000 degrees—the universe is now a thousand times bigger than it was back then. It's a linear relationship between the two. Right now the temperature of the universe is three degrees. We went from 3000 to three degrees K. That is what it is today; that's the famous microwave background. A 2000 degree object actually radiates infrared and visible light. You can see that. If it's three degrees, it's not visible to the human eye but it is to a microwave receiver. It's everywhere, so we call it the cosmic microwave background.

In the first few billion years of the cosmos, once electrons were bound to the atoms, matter was ready to do something interesting. We were ready to start building the cosmos. In those first few billion years, galaxies were manufactured. Our Milky Way was one of them. Somewhere between 50 and 100 billion galaxies were made, each containing up to 100 billion stars, each of those stars undergoing thermonuclear fusion in their cores.

Stars with more than about 10 times the mass of the Sun are very special in the evolution of the universe because they had enough mass to create enough pressure to create enough temperature to turn their cores into an element factory. They manufactured dozens of elements over their lifetime, beyond the birth elements from the Big Bang— beyond hydrogen, helium and a little bit of lithium. These stars made the elements that we are composed of. They made the carbon and the nitrogen and the oxygen and the iron. They undergo thermonuclear fusion to do that. "Thermo" means it uses heat; "nuclear" means the nucleus, and "fusion" means you're bringing them together. That's the anatomy of that word.

So what if the star just made the elements? They're no good if they're inside the star; we've got to get them out of the star somehow. It turns out; those stars explode. They blow their guts to smithereens in the galaxy—at the end of their lives—that's how they die. They're visible. We see them. We call them *supernovae*. Fortunately, their chemical enrichment, after being manufactured for the first time, is now distributed across the galaxy. In fact, the universe might view it as pollution; but, since we're made of the stuff, I think of it as enrichment.

After seven or eight billion years of this enrichment in every galaxy— but in particular the Milky Way galaxy—an undistinguished star, the Sun, was born in an undistinguished region, the Orion Arm, of an undistinguished galaxy, the Milky Way, in an undistinguished part of the universe, the Virgo supercluster of galaxies. I like to think of that as our cosmic address. It's what you would put on an envelope if you wanted to mail a letter to someone else in another galaxy, in another part of the universe.

The gas cloud that made the Sun had a big enough supply of heavy elements to spawn a system of planets, 10,000 asteroids, and trillions of comets. While most of the universe is hydrogen and helium, everything else on the periodic table is what you really need to flesh out the rest of what the universe is about. You don't make planets with just hydrogen and helium, you need the rest of the ingredients as well to make rocks and people and oceans.

Earth spent 600 million years suffering the bombardment of debris left over from the formation of the solar system, and every time an asteroid hits it heats up the area where it hits. If you were a molecule trying to become something ambitious, you'd get destroyed. If you're hit often enough, you raise the temperature of the entire Earth until you're basically sterilized. For 600 million years in the early history of the Earth, nothing could form, even though it had the native ingredients in it. It had the water, it had the organic molecules, but it couldn't make anything because it was too hot.

The vacuuming subsided. There's wasn't an unlimited amount of debris left over; 600 million years of it was enough. It began to subside. Earth's surface cooled and complex molecules began to form. What else was important to life? Liquid water was important to life. Where on Earth did that form? It didn't form too close to the Sun; otherwise all our water would have evaporated. It didn't form too far away from the Sun; otherwise all the water would have frozen. There was a nice habitable zone where the Earth was formed.

The temperature had subsided after the end of the period of bombardment and we had the right temperature—think of that as the Goldilocks effect—and the right ingredients, so that now some interesting things happened.

Within the chemically enriched liquid that was our oceans, by mechanisms still to be learned in the community of biologists, life began. From organic matter, organic non-life became organic life. That time took about 200 million years from the end of the period of heavy bombardment to the point of our oldest fossils, 3.8 billion years ago. We start out with a 4.6 billion-year-old Earth, 600 million years go by; we're at four billion years. Heavy bombardment ends, 200 million years go by and life begins.

This early life was single-celled organisms. They weren't doing calculus; they had very simple lives. There was no oxygen, so they were anaerobic. They thrived in the carbon dioxide-rich atmosphere of the Earth. As a waste product, they released oxygen. In fact, they ultimately died in their own "excrement," this oxygen. Oxygen is very bad for you if you are anaerobic; but it is very good if you are aerobic. It enabled a whole generation, a whole population of the Earth to take over that was full of aerobic creatures thriving on the oxygen made by the previous generations of single-celled organisms—oxygen in the oceans as well as in the atmosphere.

That oxygen not only enables aerobic life—that's $O_2$ in the atmosphere—there's another version of oxygen, $O_3$, that we call ozone that was formed as well. It rises up in the atmosphere and protects us from the ultraviolet radiation from the Sun. Ultraviolet photons have enough energy to break apart molecules. It's hostile to life and to the stability of life. We simultaneously get air to breathe in oxygen metabolism and protection from solar UV radiation.

We owe the remarkable diversity of life on Earth to something profound. Among all the elements that are in us there is one that is quite special, the element carbon. Carbon is plentiful in the universe, made in the crucibles in the centers of stars and spread into the galaxy. You can make more kinds of molecules with carbon than you can with all other elements combined. It is a fertile element with which to make experiments on life. When you look at the diversity of life on Earth, there's only element that could have been responsible for that, and that's carbon. We are carbon-based life, but life is fragile.

Although not as often as it once did, it still happens that asteroids come and slam into Earth, completely changing the ecosystem, if only briefly. Sixty-five million years ago, a 10-kilometer asteroid came and slammed into the tip of what is now Mexico, the Yucatan Peninsula. It made a 200-kilometer-diameter crater. If you were there, you were vaporized. If you were on the other side of the Earth, the trillions of tons of Earth's crust that was thrust into the stratosphere cloaked the Earth, blocked the sunlight, took out the base of the food chain; and if you didn't die from the impact, you died because you starved. Seventy percent of all species of life on Earth's surface went extinct at that moment, including the classical dinosaurs that we've come to know and love and respect from afar, like *T. rex*.

Some animals survived. There is a tree shrew running up and down trees and underfoot, trying to not get eaten by *T. rex*. That spawned all the mammals that followed because tree shrews survived that collision. It pried open a whole ecological niche that was not previously available to the tree shrew. They didn't have *T. rex* chasing after them, turning them into appetizers. One branch of these mammals became primates. One branch of those primates became *Homo sapiens*, these big-brained things that have what we call intelligence. That intelligence enabled *Homo sapiens* to invent methods and tools of science, to invent astrophysicists, and to deduce the origin and evolution of the universe.

Yes, the universe had a beginning. Yes, the universe continues to evolve; and yes, every one of our body's atoms is traceable to the Big Bang and to the thermonuclear fusion that went on in the center of a high mass star that gave life to us. We're not simply <u>in</u> the universe; we are part of it. We are born from it. One might even say that we've been empowered by the universe to figure it all out, and we've only just begun.

# Lecture Ten
# Forged in the Stars

This lecture highlights one of the most important discoveries in any field in the 20th century: the origin of the elements that make up life. Despite its importance, however, most people aren't aware of this discovery. Our understanding of the formation of elements did not come from the common picture that we might think of: a lone genius working night and day in a laboratory until he or she reaches a "Eureka!" moment. Instead, this discovery took place over many decades and involved many people and complicated concepts. For this reason, the discovery is hard to condense for the media and remains outside of our everyday awareness.

## Outline

I. Elements in the cosmos have two primary origins: All the hydrogen and most of the helium in the universe came from the Big Bang; the heavier elements are formed in the centers of stars.

A. As mentioned earlier, a supernova is a variety of star that has a high enough mass to create the conditions to manufacture elements.

B. If we study supernovae, we learn how they manufacture elements and distribute these elements in the galaxy. This knowledge, in turn, teaches us about the relative mix of elements found in the universe.

II. In 1957, a seminal research paper was published that brought together data from different branches of science to reveal supernovae as the primary source of the existence of heavy elements in the galaxy.

A. "The Synthesis of the Elements in Stars," was written by Margaret and Geoffrey Burbidge, along with William Fowler and Fred Hoyle.

1. For forty years before 1957, scientists had wondered whether the source of energy in the stars could be responsible for transmutation of the elements.

2. It was not until this time, however, that enough experimental data was accumulated to confirm that theory.

**B.** The results of this paper were driven by a number of "messy" questions.

   **1.** How do the various elements on the periodic table behave when subjected to extreme pressures and temperatures? This question must be answered primarily on the basis of the laws of physics. Scientists cannot do much experimenting to discover these behaviors.

   **2.** Are elements formed through fusion or fission? Which process predominates? Are these processes easy or difficult?

   **3.** Does a certain reaction produce energy (*exothermic*) or reduce energy (*endothermic*)?

   **4.** How can we explain the periodic table? Can we start with hydrogen and helium—the birth ingredients of the universe—and manufacture all the other elements?

**C.** These questions were basic in the attempt to understand the formation of the elements, but there is another question that is almost impossible to answer from first principles: What are the collision cross-sections of the nuclear pathways?

   **1.** In other words, if we want to engineer a collision between two particles, how big a target do these particles represent to each other? Are they, in fact, very small and, thus, our aim must be precise, or do they somehow take up more space and become more likely to hit each other?

   **2.** In atomic physics, researchers need to know the likelihood and rates at which elements will collide.

   **3.** The wrong cross-section calculations will result in the prediction of different kinds of reactions that would never take place.

**III.** Sir Arthur Eddington had thought about these questions at the turn of the century. His book *The Internal Constitution of the Stars* put forth the idea that the stars were the "crucibles" in which the lighter atoms were formed into more complex elements.

   **A.** Eddington knew that an exotic environment with very high temperatures was required to force light elements to collide and make heavier elements. However, he was missing a piece of physics needed to figure out the whole story.

**B.** In the 1920s, with the development of quantum mechanics, the missing piece was supplied.

 1. If we consider bringing together two hydrogen nuclei, we know that their two protons will repel each other under normal circumstances. We need to heat the gas so that the particles will move faster and get closer together. If we can get these particles close enough, the strong nuclear force will take over and they will bind.

 2. The particles must cross the *potential barrier*, that is, the electromagnetic resistance of the two positive charges.

**C.** Eddington calculated the temperature of the center of a star at about 10 million degrees. Then, he calculated the temperature required for protons to collide with enough speed to bind together. That temperature is about a billion degrees.

 1. It seems obvious that elements could not be formed in stars at 10 million degrees, but we knew that the hottest place in the galaxy was the center of a star. If elements weren't formed there, then where else?

 2. We now know that the speed of protons at 10 million degrees is sufficient to make them collide, but not in the traditional way that we might think.

**D.** The collision of protons at 10 million degrees is possible through a phenomenon in quantum mechanics that has no analog in everyday life—*tunneling*.

 1. Think of rolling a toy truck up a hill. Under normal conditions, you would have to roll the truck up the hill at a certain speed for it to reach the crest and go over.

 2. In quantum mechanics, under certain conditions of temperature and pressure, if you roll the truck up the hill, a tunnel will open in the hill and the truck will pass through to the other side.

**E.** Thus, in trying to overcome their repulsion, some protons do connect at temperatures as low as 10 million degrees. Before quantum mechanics, however, we had no way to know this.

**IV.** Next we might ask about the relative amounts and distributions of elements in the universe.

**A.** In 1931, Robert Atkinson published a paper about what goes on inside of stars. He tried to show how the elements were built up one by one and how this process accounted for their relative distributions.

**B.** Once again, Atkinson did not get the full answer because he was missing a piece of the puzzle—the neutron—which was discovered a year later by James Chadwick.

**C.** Because neutrons have no charge, they can be used to build up the count of particles inside the nucleus. There is no resistance to the addition of neutrons and no change in the species of the nucleus when neutrons are added. In other words, a hydrogen atom is still hydrogen with the addition of neutrons, but it becomes an *isotope*, an element with varying numbers of neutrons and the same number of protons.

**D.** Some elements reject the addition of extra neutrons. In an atom of such an element, the added neutron is spontaneously transformed into a proton that releases an electron. A new proton, then, has joined the nucleus. This process is an effective way of building up elements.

**E.** This process varies in different elements, but it leads to an understanding of the formation and distribution of elements in the universe.

**F.** The final piece of this puzzle was the collision cross-sections, most of which came out of the research that went into the Manhattan Project. We now knew how elements were made in the centers of high-mass stars.

**V.** How are these elements distributed in the universe?

**A.** Thermonuclear fusion takes place in the cores of high-mass stars. Such a star begins by fusing hydrogen into helium, which results in a loss of mass. The mass is converted, according to $E=mc^2$, into an enormous amount of energy.

**B.** This process continues as the star fuses helium into carbon, carbon into oxygen, oxygen into neon, and so on. The star is receiving energy at every phase, but it is not as efficient; it begins to burn through these heavier elements very quickly.

**C.** As the fusion process moves up the periodic table, it reaches iron. The star will collapse a little bit to increase its temperature and begin to fuse iron. Extremely high temperatures are required for this fusion.

**D.** The fusion of iron absorbs energy—it doesn't release energy. However, the creation of energy supports the star from collapse; without a source of energy, the star has nothing to hold itself up. It destabilizes and collapses in a matter of hours.

**E.** The collapse of the star rebounds from the center in a titanic explosion that we call supernovae. These are visible across the galaxy and spew a trove of enriched chemical elements throughout the universe.

**F.** The remnant of one such supernova is the Crab Nebula, a star that exploded and was recorded by the Chinese on July 4, 1054 A.D.

**VI.** The 1957 paper of Burbidge, Burbidge, Fowler, and Hoyle combined the tenets of quantum mechanics, the physics of explosions, the latest collision cross-sections of atomic nuclei and resulting nuclear pathways, and basic stellar evolutionary theory to account for the existence, distribution, and relative abundance of elements in the cosmos. Noting that we—human beings—are made of stardust might summarize their conclusions.

**Suggested Reading:**

Altshuler, Daniel R. *Children of the Stars*. Cambridge: Cambridge University Press, 2002.

Spitzer, Lyman. *Searching Between the Stars*. New Haven: Yale University, 1982.

**Questions to Consider:**

1. Where do all the heavy elements in the universe come?

2. Describe two reasons why carbon is so useful to life.

# Lecture Ten—Transcript
## Forged in the Stars

Welcome back to My Favorite Universe. For this lecture I'm going to highlight what is, in my judgment, one of the most important discoveries of the 20th century. It relates to something that people have been thinking about across time and across culture. If you read through texts, whether religious texts or scientific texts—it won't matter—look through the history of time at a culture's writings and in there you will find that people have looked up at their night sky and wondered what is their place in the cosmos. Where did it all begin? Where is it all going to end? These are common themes. Maybe those questions are genetically encoded within us.

This particular lecture entitled "Forged in the Stars" is a discussion of what is the origin of the elements that make up the human body, that make up Earth itself and everything that is not just hydrogen and helium in the cosmos. Most people don't know the origin of the elements. Now how is that possible, when I just told you that it is, in my judgment, the most important discovery in any field in the twentieth century? Most likely you don't know about it because it was not your classical kind of discovery—where you have the lone genius burning the midnight oil saying "Eureka!" in the wee hours of the morning, and then being rushed by the press to find out what the discovery was and then put headlines the next day. That's not how this discovery happened. It's not a very media-friendly story in spite of its importance.

This particular discovery took many decades and many people. It uses complicated math, and it's very hard to create sound bites out of it. This was the nature of the discovery of the origin of the elements, and that is the subject of this lecture.

I could tell you now the elements in the cosmos have two primary origins. One of them is the Big Bang itself. We got all of our hydrogen and most of our helium from the Big Bang. Everything else came from stars, the crucible in the center of stars—this hot place—and in particular from one variety of star that has very high mass. It's a star that ultimately explodes. We call them *supernovae*, and they've been mentioned several times in previous lectures. If you study how supernovae manufacture elements, and how they explode and distribute them in the galaxy, you learn not only that such elements exist, but you also get the relative mix of the elements that we find in the universe, and you find out where they are distributed in the galaxy. You get all that for free once you understand how supernovae work.

We trace that knowledge to a seminal research paper that came out in 1957. It was published in the *Reviews of Modern Physics*, a respected journal. It was titled, "The Synthesis of the Elements in Stars." How much more simple can you get? The authors were Margaret Burbidge and Geoff Burbidge—they were a husband and wife team—along with William Fowler and the inimitable Fred Hoyle himself. Fred Hoyle is a brilliant man known for his extraordinary ideas that are occasionally correct; and, as a participant in this paper, this was one of them, one of his crazy, brilliant, correct ideas.

In the 40 years prior to 1957 people had mused about whether the source of energy in the stars could be responsible for the transmutation of elements. People had wondered, but they had fragments of knowledge with which to make their picture, so there was a limit as to how far they could go. Once enough time had passed and enough experimental data had accumulated, Burbidge, Burbidge, Fowler, and Hoyle brought together all of those pieces and implicated supernovae as the primary source of the existence of heavy elements in the galaxy. This is what made that paper significant, but it was many decades in the making.

There were messy questions that led up to that, such as how did the various elements on the periodic table of elements behave when subjected to extreme pressures and temperature? You can't go visit the center of the Sun and bring out your measuring devices; you'd be vaporized long before you got anywhere near it. You have to be cleverer than that. You have to trust your knowledge of your laws of physics. You can do some experimenting, but that's about it. When you try to make the elements, do you fuse them together or do big elements only break apart? Do you use fusion or fission? What's the predominant process? How easily is this accomplished?

These were questions that were driving the results of that paper. When you have a reaction, does it give you energy, is it exothermic, or does it take energy away—endothermic? That's a really important question with regard to the survival of a star because stars are in the business of making energy.

Not only that, can you explain the periodic table of elements itself? It's more than just a chart of boxes sitting in front of your chemistry class. There's a lot of action going on with this stuff. Can you start with hydrogen and helium, the birth ingredients of the universe, and manufacture all the other elements on that periodic table? Is that possible? Is that sufficient to synthesize everything on the table? If you could do that, that would be kind of like modern alchemy.

You read about the Merlin characters in their labs with test tubes trying to turn base metals into gold. We can do that today. We know how to do that. The reason why they couldn't do it is because they were experimenting in the domain of electromagnetism. They didn't know that they had to reach the domain of another force in order to transmutate the elements. They had to get into the nucleus of the atom and electromagneticism doesn't go there; the strong nuclear force does. Had they known about the strong nuclear force 500 years ago, they might have known how to get into the atom and split it and make two other kinds of atoms.

That's the sort of the basic questions and a basic attempt to understand the formation of the elements; but there is another problem, a very real problem that it's hard or even impossible to deduce from first principles: What are the collision cross-sections of the nuclear pathways? That's fancy talk but it's not anything deeper than if you have two particles—in order to slam them together—how big a target are they to each other? Are they small so that your aim has to be really good, or do they somehow take up more space so that when you slam them together, they're very likely to hit each other? We call that a *collision cross-section*. You have an intuitive sense of this because, when you're on a highway and see a doublewide house coming—that has a large cross-section—you take tremendous precautions so that nothing collides with it. In atomic physics you want to know what that cross-section is. You want to know it exactly so that when you start bringing your ingredients together, you're going to know who's going to hit who, at what rate, and who's going to do it the best, or the slowest or the fastest. From that, you derive the nuclear pathways of what goes on in your soup of atoms.

If you get the wrong cross-sections, you have no idea what's going on. You might as well just give up and go home because one wrong cross-section in your calculation will give you a picture that will spawn all kinds of other reactions that would have never happened in the first place. It would be like going into the New York City subway system but with a map of the London subway, trying to navigate through the New York City system. That's just not going to work. Collision cross-sections are important to learn and important to know. This whole effort is difficult and, like I said, it did not lend itself to sound bites.

Others had thought about the general problem. One of the most brilliant astrophysicists of the turn of the century, Sir Arthur Eddington, wrote a seminal book titled *The Internal Constitution of the Stars*. I quote from that book, "I think that the suspicion has been generally entertained that the stars are the crucibles in which the lighter atoms, which abound in the nebulae, are compounded into more complex elements." People knew the universe had larger elements. They knew, but where did they come from? Was the universe born that way? This is way before the Big Bang, so they thought either somehow the universe happened that way or they're made. He suspected that they were made. In 1920 atomic physics was going strong. There were the beginnings of nuclear physics, but it was the dawn of quantum mechanics. The full understanding of particles and atoms and nuclei was still to come, but he suspected that they were on the right track.

You could ask the question, "Where else in the entire cosmos would you make heavy elements, but for the centers of stars?" Where are you going to find an environment to slam light elements together to make heavy elements? Where is that going to happen? It's not going to happen in your back yard. It's not going to happen in your kitchen. You need something as hot as you can imagine, and that's the centers of stars. That was well known at the time.

The real solution required the discovery of quantum mechanics in the 1920s. There's no way Eddington could have figured out the whole story. It was missing physics. Physics had to be invented to complete that story. Quantum mechanics, as I've alluded to many times, is the science of very small things: atoms and molecules. Something to consider is protons. A hydrogen nucleus is a single proton, the lightest, simplest element. If you want to bring two hydrogen nuclei together, they're both positively charged and they're going to repel. So what do you do about it? Slam them together faster and they'll get a little closer before they repel. How do you make it go faster?

One way to do it is to heat the soup. In heated gases the particles move faster and faster and faster as the temperature rises. If you get them close enough together so that they're within the domain of the strong nuclear force, the strong force kicks in and binds the two protons together. They had to cross this what we call *potential barrier*, though—this electromagnetic resistance of the two positive charges. They had to overcome that. As I said in one earlier lecture, it is like taking a toy truck and rolling it up a hill. If you don't give it enough energy it sort of rolls back, and you roll it faster and it gets higher up the hill and rolls back. There is a speed at which you can roll it to make it get all the way to the top and then it goes over. That's what's going on with these protons

Eddington is not stupid. Eddington calculated the temperature at the center of the star. He got that right. You don't need quantum mechanics for that. He said it is 10 million degrees or so. Then he calculated the speed with which protons must approach each other in order to collide. Do you know what temperature he got? He got a billion degrees. That worried him because the two temperatures had no correspondence with each other. They were vastly different temperatures. If you want to rely on collisions in order to give you a heavier element, you might be telling yourself at this point that the stars are not the place, as hot as they are, it's not going to work. Tens of millions of degrees is far away from a billion degrees.

That's why there was as much resistance as there was at that time for considering stars to be the place where heavy elements would be formed. But there were the diehards who said, "I don't care what you say, all I know is the hottest place I can think of in the whole galaxy is the center of a star; so if you want interesting, atomic things to happen, nuclear things to happen, if we don't look there I don't know where else to look." That's the right attitude, because it turns out that the speed of protons at 10 million degrees is enough to make them collide, but not by the traditional way you might think.

There is a phenomenon in quantum mechanics that has no analog in everyday life. It's called *tunneling*. Let's go back to my little truck and I'm trying to roll it up the hill. I try to shove it up the hill and it rolls back because there's not quite enough energy. In quantum mechanics, depending on the ambient temperature and pressure and conditions, one of these times I roll the truck, a hole opens up through the mountain and it goes through the mountain and comes out the other side. It gets to the other side of that mountain for free. That would be pretty interesting if that happened in everyday life, but it doesn't. It happens in the realm of the atom. Some of these protons, which are trying to overcome their repulsion, actually do connect at temperatures as low as tens of millions of degrees. You didn't need the billion degrees. But, before quantum mechanics was discovered, there was no way to think about that problem. There was no way to answer that problem, no solution waiting for you.

Quantum mechanics tells us the temperature is hot enough to make the elements. That's good, but what about the relative amount of each element? That's another kind of question to ask of the star. The 1920s are over and quantum mechanics is in the can. In 1931 there was an astrophysicist named Robert Atkinson who published an extensive paper about what was going on inside of stars. His abstract summarizes what he was after: "Synthesis theory of stellar energy and of the origin of the elements in which the various chemical elements are built up, step by step from the lighter ones in stellar interiors by the successive incorporation of protons and electrons, one at a time."

We already accounted for the existence of elements and he is trying to find a way where, one by one, you build up the elements. In his paper he talks about what kind of relative populations of the elements you would get, but he didn't have a chance of getting the right answer. He was missing a particle. He had the right physics, but he was missing a particle. A year later in 1932 James Chadwick discovered the neutron. You can't talk about nuclear physics without the neutron. Neutrons go hand in hand with protons in the nucleus. That's an incomplete theory. You left the salt out of the soup. No matter how smart you are; you're not going to get the right answer without the neutron.

The fun part about neutrons is they don't have a charge. So here's a nucleus with a fat positive charge and you take a neutron and toss it toward the nucleus, it just walks in the front door. There's nothing to resist it. You can have a nucleus, feed it neutrons and build up the count of particles inside the nucleus. If you hand a neutron to the nucleus, it doesn't change the "species," it's still hydrogen if you add a neutron to it. We have a term for it. It's called an *isotope*. The isotope of an element has a varying number of neutrons and keeps the number of protons. There are some elements, if given an extra neutron, won't like the extra neutron and spontaneously—again we learn this from quantum mechanics—spontaneously, the neutron becomes a proton and releases an electron. The charge is cancelled between a proton and an electron—as they are neutral in the neutron in the first place—so all the charges work out. Almost for free a proton joined the nucleus.

This process is a very effective way of building up elements. It is true that with some elements, if you give them one neutron they go unstable and they kick the whole neutron out again. But if they have two neutrons, they are stable and given time they will convert the neutrons to protons. Think about it. If you have a weak flux of neutrons into your atoms and into your nuclei, an atom that is unstable with a neutron is not going to do anything with it. If it's stable with the neutron, it will mutate it into the next element up on the periodic table.

If you have a heavy stream of neutrons, one can be added which makes the atom unstable; but before that neutron can be released, another one comes in. Now there are two. The nuclear physics of it says that this atom is stable with two even when it was not stable with one. So with two neutrons it's stable. Now it can undergo a decay again and turn a neutron into a proton and an electron. One method is called *slow neutron capture*; the other is called *fast neutron capture*. The only lesson is that now there are more ways to make elements. It is not simply slamming protons together the macho way; there are also less spectacular ways of building up the population of elements on the periodic table.

When you combine all of those features, each one of them gives you a different mix of different elements. Put them all together and you begin to recover the exotic distribution of elements in the cosmos. You begin to discover why it is, for example, that on average all elements with an even number of protons are more abundant than the adjacent element that has an odd number of protons. That's odd. Those elements needed two particles to come in. If you're actually slamming a helium nucleus in, there are two protons right off the bat and it grows just by accumulating helium nuclei. You're hopping into pairs. If you look at the abundance on the chart you see this fascinating distribution, and it all comes out of this analysis.

What about those collision cross-sections? Those are hard. We got our collision cross-sections to feed the research that went into the 1957 paper by Burbidge, Burbidge, Fowler and Hoyle from the research that went into the Manhattan Project, the American program to build the first atomic bomb in the Second World War. When you're at war, a lot of money gets spent. A lot of effort gets invested, not because you're curious about the science, but because a weapon is being made. Your defense is of paramount importance. A sidelight of that fact is, we got from that research—once it had become declassified—collisional cross-sections of particles moving in and out of an atomic nucleus and it enabled those calculations in the first place.

Now we know how to make the elements: in the centers of stars and in particular in the centers of high-mass stars that have enough temperature and enough pressure to keep that going right on up the periodic table of elements. Let's begin by asserting something that is not a strange concept: stars are in the business of making energy and that's all they know how to do. If you look at a cross-section of the Sun what you see is down in the center is the hottest place—it's much, much cooler out to the surface—in the core is where all the thermonuclear fusion is going on. The Sun is making heavier elements. It's converting hydrogen into helium, but that's not the interesting elements. The interesting ones are the ones that we're made of, and who's going to make those? Those are the high-mass stars. They start out making hydrogen into helium, fusing the two—fusing hydrogen, becoming helium. There's a loss of mass there if you do the arithmetic. You start out with more mass than you end up with. What happened to the mass? It all got converted according the $E=mc^2$ and you get an enormous amount of energy from that lost mass.

The process continues as high enough mass stars convert helium to carbon, taking three helium atoms and getting carbon out of them. If you have high enough mass it will keep cranking this through. It runs out of one element and starts manufacturing the next. It goes from helium to carbon, carbon to oxygen, oxygen to neon and this continues; and it's getting energy at every phase except it's not as efficient. It blows through those heavier elements very quickly compared with how much time it spent converting hydrogen to helium.

As it moves up the periodic table, one of the elements it's going to end up with is iron. Time to fuse iron. The star will collapse a little bit, the temperature will rise and it will begin to fuse iron at extremely high temperatures, much higher than what the Sun is doing right now or the center of a star while it's converting hydrogen to helium. As it starts fusing iron there is a problem. If you fuse iron, it absorbs energy. It doesn't release energy. That's a bad day for the star because a star only knows the creation of energy. It's that creation of energy that supports it from collapse.

Without a source of energy, it is hosed. The star has nothing to hold itself up. If it gets to iron and fuses iron, not only is it not good for it because it's not giving up any energy, it is actually absorbing energy and the whole star destabilizes and it collapses in a matter of hours. The collapse of that star rebounds from the center in a titanic explosion that is as luminous as a billion stars. We call those objects *supernovae*. They're visible across the galaxy, across the universe. Looking at an image of a galaxy half way across the cosmos, you see a little smudge there but there's a bright light sticking on one of the spiral arms of the galaxy. That's a supernova that just went off. These things happen in real time when they happen. You can watch them get bright and watch them get dimmer as the explosion runs its course.

This is such an active place that there are a lot of free neutrons and protons running around. Not only did it manufacture elements on the way to iron, there is a lot of manufacturing going on after iron as well, slow neutron capture, fast neutron capture, a little bit of extra fusion on the side. It is a trove of enriched chemical elements.

One such remnant of a supernova is the Crab Nebula. The Crab Nebula is the remnant of a star that exploded and we think was seen by everyone on Earth; but it was only recorded by the Chinese on July 4, 1054 A.D. The Chinese records say there's a new star in the sky. We now know it was that supernova making this residue—this 2000-year-old explosive remains—which we have poetically referred to as the Crab Nebula. You notice on it there are fibrous parts to it. Each one of those areas contains a strand of heavy elements lifted out of the center of the star and brought into the galaxy for the rest of the galaxy to share. So that subsequent stars that are made can make something other than the central star, like planets and even people.

The next image is of another supernova remnant, Cassiopeia-A we happen to call it. It's in the constellation Cassiopeia. You will notice once again how explosively thrust forth these gas clouds are. If you analyze the chemical composition of each of those strands, you'll find very high iron, and oxygen, and carbon, and all the elements we've come to know and love as living beings that thrive on the heavy elements of the cosmos.

What Burbidge, Burbidge, Fowler and Hoyle did was take the well-tested tenets of quantum mechanics and combined it together with the physics of explosions and the latest collisional cross-sections for the atomic nuclei. They also looked at all the nuclear pathways that can result, based on those input parameters, and combined that with basic stellar evolution theory. When you've done all of that and you've done your homework, you account not only for the existence, but also for the distribution and relative abundance of elements in the cosmos. It was in 1957 where something was determined about us that I will never forget, and I will tell as many people as I possibly can. You might have expected it all along but it remains true, not just figuratively, but literally—that we are all stardust.

# Lecture Eleven
# The Search for Planets

**Scope:**

It seems fairly obvious that planets outside our solar system would be found in the Milky Way, but before the 1990s, our planets were the only ones known in the galaxy. Now, we know about at least 100 planets outside the bounds of our solar system. Astrophysicists did not look for these planets just out of curiosity but as part of a systematic search for other planets that could support life. This lecture discusses the tools and methods used to find other places in the universe that might be hospitable to human life.

## Outline

I.  Given that our Sun is just an "ordinary" star, and it has enabled the life of our species, we might wonder if other "ordinary" stars can also support life.

    **A.** We can find the total number of possible stars that might support life by multiplying the number of stars in a galaxy (100 billion) by the number of galaxies in the universe (50–100 billion). The result is one sextillion, or a one followed by 21 zeroes.

    **B.** That number is 100,000 times larger than the total number of sounds uttered by all human beings who have ever lived.

    **C.** Each of those stars could have at least one orbiting planet that is capable of supporting life.

II. How do we go about finding these planets?

    **A.** We can't just look up in the sky and find a likely planet, because a planet is usually in orbit around a host star that is much greater in brightness. A planet might be 1/100,000,000 the brightness of its host star. Astrophysicists needed to find some methods to block the light of the host stars.

    **B.** Many planets, as well as solar systems in formation, give off principally infrared light as opposed to visible light. Using an infrared telescope, we might increase the brightness of a planet to 1/10,000,000 that of its host star. Within the optics of the telescope, an eclipsing disk can also be added to block the light of the host star.

**C.** Using this method, researchers in the 1980s discovered an orbiting planet in a solar system in formation around the star Vega, one of the brightest stars in the night sky.

**D.** Close examination of Vega revealed infrared emission surrounding the star. Even closer analysis enabled us to see craggy chunks of rock and dust still in the act of forming a system of planets.

**III.** To find planets that are capable of supporting life, we need to look for Sun-like stars and stars that are not in multiple-star systems.

    **A.** Most stars in the sky are in multiple-star systems, but planets in orbit are not stable in the changing gravitational field of these systems. These planets are often jettisoned into interstellar space and become *vagabond planets*. We need to look for single stars with stable orbits.

    **B.** We might have many views in the galaxy of a single host star with a planetary system. For example, we might be able to look down on such a system for a bird's eye view. What we need to find, however, is an edge view.

        **1.** In an edge view, a large planet, similar to Jupiter, for example, will orbit and move in front of the host star. This orbit will not totally eclipse the star, but it will block some of the star's light.

        **2.** We can monitor that star's light output. If it is normally stable but is blocked with some regularity, we know that a planet is in orbit around the star. In other words, we look for regularly occurring variations in the light of an object that might otherwise be stable.

        **3.** We need to be careful in drawing conclusions from this method of observation because there is a possibility that the star's light is being blocked by another phenomenon, such as sunspots.

**IV.** Another method for finding planets is called *microlensing*, which taps into the theory of general relativity.

    **A.** General relativity states that the fabric of space and time warps in the vicinity of a source of gravity.

    **B.** A vagabond planet moving in front of a star can have an effect on the star other than blocking its light. The gentle curvature of space around the planet may be just enough to bend and focus the paths of light emanating from the star from different directions in the cosmos. As the planet drifts across our view, the starlight greatly increases in brightness, and then drops off.

C. The planet's gravity serves as a lens to focus the light of a distant star behind it.

D. This method of finding planets has its drawbacks. First, it is non-repeating; the planet drifts past the star only once. Second, the likelihood of a planet crossing the view of a star is exceedingly low.

V. A tried-and-true method for discovering planets is by using the Doppler effect, discovered by Christian Johann Doppler, a German physicist, in the 19th century.

A. Doppler noticed the difference in pitch of a train whistle as a train approached, then receded. He found the mathematical formula that calculated the rate that the frequency changes. This frequency change is a basic property of sound waves emanating from an object in motion; the same property also holds for light waves.

B. We usually think of planets orbiting around a star in the center, but this view is not completely realistic. In fact, the star and the planet are both orbiting around their common center of gravity, which is generally not at the center of the star.

C. Thus, we can measure the change in the frequency of the light as the star approaches and recedes. The change in frequency is the same as the change in wavelength.

D. In other words, the host star is moving in reaction to the orbit of the planet, and the movement of the star has the same period as the movement of the planet. The amount of the star's motion is also a function of the mass of the planet. The bigger the mass of the planet is, the greater the motion of the host star.

E. Using this method, we can also find not just one orbiting planet but also multiple planets. We can see more than one level of variation in the movement of the host star.

**F.** The first planets discovered using this method were the size of Jupiter. Obviously, we would expect to find large planets having a greater effect on the movement of the host as opposed to small planets, such as Earth, which would have a smaller effect on the host.

   **1.** These large planets were also orbiting very close to their host stars, where we would never have imagined we would find them.

   **2.** Our assumption was that rocky planets would be close to the host, while massive gaseous planets would be farther out, similar to our solar system. Thus far, we have not seen that kind of system and have had to change our theories of planet formation.

   **3.** Our methods may be preventing us from seeing other planetary systems accurately. It is possible, for example, that we have not been using this method long enough yet to establish a baseline to measure a full orbit of another solar system that might have a very distant, massive planet. Such a planet may take eight–20 years to complete an orbit, and we have been using this method for only about 10 years.

   **4.** In a star system in formation, a massive planet may move close to the host star by flinging debris out of its way as it migrates in.

**VI.** Will we ever find any planets to visit?

   **A.** If we were able to travel 1000 miles per second—100 times faster than any human has ever traveled—it would take us 800 years to reach the closest star to the Sun, which doesn't have any known planets. We would need 2000 years to reach a star that we know has planets.

   **B.** NASA has plans to build a Terrestrial Planet Finder (TPF) and combine it with the Space Interferometry Mission (SIM). The result will be a bank of telescope dishes working in harmony to provide the sharpest possible images of the smallest possible objects in the universe.

   **C.** What we want to see is an Earth-like planet, its oceans, atmosphere, and so on. We would use spectrographs to measure the chemistry of such a planet's atmosphere, which is a product of what's happening on the surface.

1. The discovery of oxygen in a planet's atmosphere, for example, would be evidence of metabolism on the planet.

2. A spectrograph might also detect CFCs, the destroyers of ozone, or hydrocarbon contaminants, the product of global deforestation. Such discoveries would be sure signs that the planet does not have intelligent life.

**Suggested Reading:**

Dorminey, Bruce. *Distant Wanderers: The Search for Planets Beyond the Solar System.* New York: Springer-Verlag, 2001.

**Questions to Consider:**

1. What is the Doppler shift, and why is it so useful for discovering exo-solar planets?

2. Among the first hundred solar systems discovered, how do they compare with our own? How are they similar? How are they different?

# Lecture Eleven—Transcript
## The Search for Planets

Welcome back to My Favorite Universe. This lecture, Lecture Eleven, is entitled "The Search for Planets". The search for planets takes on many dimensions. You might think, why search at all. We're pretty sure our galaxy, the Milky Way galaxy, has planets in it because the sun has eight or nine planets. Before the 1990s, our eight or nine planets were the only known planets in the galaxy. Right now we happen to be rising through 100, 100 planets known outside of the bounds of the solar system. Those planets were not discovered just because people were curious whether there were planets elsewhere in the galaxy. I always had the confidence that if we looked we would find them. No problem there.

What's really going on here is that we're trying to find worlds that could support life and if the Sun is an ordinary star: not too bright, not too dim, not too hot, not too cold and not too massive. If it's ordinary and it has all these planets—and one of them has life on it— then imagine other ordinary stars in the galaxy. It becomes an exercise of not just simply filling catalogs of planets as we've done before, searching for particular types of stars, even particular types of galaxies. We're looking for places that could have life. Suppose for every ordinary star, no matter how many planets it had, one of those planets was capable of supporting life, as we know it. How many planets might that be in the whole galaxy? How about the whole universe?

There are 100 billion stars in the galaxy, and in the whole universe there's about 50 to 100 billion galaxies. You take those two numbers and multiply them and that gives you the total number of ordinary stars. That's a one followed by 21 zeros, one sextillion. Most people have never had an occasion to greet that number, one sextillion. Let me give you an idea of how big it is. It's 100,000 times bigger than the total number of all the sounds ever uttered by all human beings who have ever lived. That's how big that number is. That's, how many stars there are in the universe; and perhaps, that's how many planets there are, which are capable of supporting life—unless of course, we're unusual. That's an unpleasant thought. I'd like to think that we're common.

Let's take a look to get a sense of how many stars there are in this view of a tiny piece of the sky pointed toward the center of the Milky Way galaxy, toward the constellation Sagittarius. Sagittarius gets the award for having the most stars in it because the stars that trace out the constellation Sagittarius are sitting in front of our sightline that goes straight to the center of the galaxy and all of these stars are crammed in there. That's just stars in one patch of sky in our own galaxy. Imagine summing that up for the whole galaxy and across the universe. Dare we think that we're alone?

Theorizing about there being other places where you might find life wasn't always met with support or praise. In the year 1600, Brother Giordano Bruno—who was a monk—lost his life. How did he do that? It wasn't really his fault. In the Square of Flowers in Rome, the Catholic Church burned him, naked, at the stake. Had he been just an ordinary heretic they would have just burned him without removing his clothes, but he was an impertinent heretic. Why? What was his crime? He suggested that the universe must be infinite because, if it were not infinite, then it would have to be here instead of there; and how could it be in one place and not another? If it's finite, it's leaving out some space over here. His philosophical sensibility told him that the universe must be infinite, because it can't be one place and not another, number one. Number two: From the vastness of this infinite universe, he concluded there must be many worlds beyond Earth just like Earth—and he was rather adamant about this. This got him into deep trouble, enough trouble for him to be burned at the stake. It was against scripture, against Catholic interpretations of scripture. There's a statue to him in the Square of Flowers in Rome. He's posed with his robe on. My last time in Rome I had a pilgrimage to that statue, thinking to myself, they don't do that any more—fortunately—because I'm talking about planets beyond Earth all the time; and not only is that work praised but it sometimes makes headlines.

How do we find these planets? It's a challenge. You don't just look up and say, "Oh, there's a planet." It takes much more than that, because these planets are in orbit around a host star that is vastly greater in brightness than the planet itself. There's a challenge there. You have to detect something that may be only one, one hundred-millionth (1/100,000,000) the brightness of the host star. It would be like catching a firefly and putting it in one of those Hollywood searchlight beams they use for movie openings. Just toss the firefly into that beam, step back and say, "Where's the firefly? Show me the light of the firefly." It's there, but you have to somehow block the rest of the light, eclipse it, occult it; somehow suppress it so the light of the firefly can show up a little better.

Many planets, as well as solar systems in formation, give off principally infrared light—as opposed to visible light. All they do is reflect visible light, but they give off infrared on their own. They are brighter, compared to the host star, in the infrared than they are in the visible. What you might be able to do is to whip out the infrared telescope and in that way the contrast is not 100 million. You might gain by a factor of 10. Maybe it's only 1/10,000,000 the brightness instead of 1/100,000,000. These are tactics we use. They're still crude, but they're tactics nonetheless.

In combination with that, we've found another way. I don't know if you've ever seen a total solar eclipse. If you go see one, you're out there, the Sun is there, the Moon goes in front, and right when the Moon fully covers it, the sky goes dark and stars come out. You have blocked the light of the host star, enabling you to see things that are much dimmer that were always there. You just didn't notice them before because things were too bright.

We have figured out a way, within the optics of the telescope, to put an eclipsing disk in front of the host star, blocking out most of its light and enabling us to see whether there is any signature of a tiny bit of light nearby, possibly in orbit around the host star. You hope that     it's not just a chance juxtaposition of another star in the background. That would just be bad luck.

One of the first things discovered by this method was one of these orbiting disks of a solar system in formation, and that was the star Vega, one of the brightest stars in the nighttime sky. It's visible throughout the evening sky in the summer or very early hours, the wee hours of the morning in the winter. Vega goes right overhead at night for most residents of North America; and close examination of that star revealed that there was sort an infrared emission surrounding the star. A closer analysis revealed it to be craggy chunks of rock and dust still in the act of forming a system of planets that we have come to take for granted that exist in our own solar system.

This fact about Vega was well known to the authors of the story "Contact." You might remember the film where they chronicled that first encounter with extraterrestrial intelligence. The signals were coming from Vega, and the scientists upon realizing this said, "They can't be coming from Vega." You can't have a civilization on Vega because there are still rocks trying to form a planet. If you don't have a planet, how are you going to have a civilization? This was an argument to wonder further, how it was you would have signals coming from such a planet. I was proud of the producers and the authors for having included late-breaking scientific news in the storyline of a science fiction film. They could have just made stuff up; it is science fiction after all. But they didn't. They kept with the facts. Later on it was revealed that the alien intelligent civilization evolved somewhere else and had an outpost on Vega. We were all excited. This happened back in the 1980s when this disk of material was found around Vega; but again, what we want is planets with stuff on them.

Where should we look? There are a lot of stars out there, a lot of stars in our Sagittarius star field. Should we start in one corner of the photo and just sort of work our way? We can do a little better than that. We can do better because we know that the Sun has planets. Why not look for other Sun-like stars? That way we'll look for life as we might expect it to be.

In that search we'd better remove from that sample those stars that are in binary and triple and multiple star systems. On any night, if you go out with binoculars or even a modestly powered telescope and look up, half of the dots of light you see when you zoom in on them—you'll see that they're actually two stars or three stars. If you waited long enough, thousands of years or so, you'll see them in orbit around each other. These are multiple star systems. Planet orbits are not especially stable if they're trying to navigate their way around a changing gravitational field. Often they get jettisoned into interstellar space. We expect a whole population of vagabond planets to be roaming the galaxy having been jack-in-the-box jettisoned from their original star system. You're not going to look to the multiple star systems. You want to get the single stars where they have stable orbits.

Maybe there are interesting planets around high-mass stars that are extremely luminous, or low-mass stars that are relatively cool and dim. Maybe, but I don't want to start my search there. I have limited research funds. I want to maximize my chance that I'm going to find what I'm looking for because I know that we exist. We already have a tested concept; it's us. Later on we'll go back to the other stars— once we're happy with looking for ourselves then we can look for the weird cases. I don't have a problem with that, but first things first.

How do you find them? Is there a better way than squinting and blotting out the star? There are several ways. One way that has recently been pioneered, because we really didn't think it was possible: You have the host star and let's say it's got its planetary system. There are many views of that planetary system. You might have a bird's eye view—birds don't fly in space of course so I'm being figurative. They would just drop. There's no air. You have your view from the top.

There are a lot of views of the planetary system that you might have, but one of them might be edge on. In all the random configurations of their orbital plane, there might be a system out there with one that is edge on. If it's edge on, and there's a nice, big, fat planet there like a Jupiter. That Jupiter is going to swing around and move in front of the host star. It's not going to cover the whole star like a Sun/Moon eclipse, but it's going to block out some of the light of that star. If that star has normally stable light output, and I'm monitoring its output, and all of a sudden there's a blockage, and it repeats at regular intervals, I've got an eclipse on my hands. I've got light being blocked from that host star by something that passes in front of it at regular intervals. I've got a planet. Planet eclipses is a major frontier that we now have, because one of the easier things to do is to check for the variation in light of an object that might otherwise be stable in its light output. Legions of amateur astronomers can help out with this effort in their backyards.

What you hope is going on is that the planet of the host star isn't very pockmarked. You may remember or you've heard that the sun has these things called *sunspots*. They're spots on the Sun—this is astrophysics; we call it as we see it. They're sunspots. We don't invent Greek and Latin names for these things. The Sun has sunspots and sometimes it has more than other times. As they move in front of the Sun, their luminosity—the radiant energy from the Sun— changes. We have to be very careful that we're not measuring just some rotation of the Sun carrying a blot, a dark region in front at regular intervals. We have to make sure we're not looking at that rather than a planet that would come in front of it, blocking its light. There's not an obvious answer as to how to protect against being fooled. You know that a planet would eclipse at regular intervals basically forever, while sunspots come and go. You're going to check for signatures that might disappear, so either the planet got eaten or the false signatures were sunspots.

Another way to look is something called *microlensing*. This is a brilliant, brilliant, high-tech method. It taps some of the basic tenets of general relativity—Einstein's general theory of relativity which asserts, and has been demonstrated to be true—that, in the vicinity of a source of gravity, the fabric of space and time warps. It doesn't matter how big the gravity is. It warps much more around high sources of gravity, but around low sources of gravity it warps just a little bit.

If there's a star in the distant part of the galaxy and I have a planet, one of these vagabond planets—it wouldn't have to be a vagabond, but let's just say a vagabond planet—if that vagabond planet passes in front of the star, which is so far away, it's not going to actually block much of the light because it's far away and the planet is just this infinitesimally small dot moving across the surface. That's not what we're measuring here. The gentle curvature of space around the planet is just enough so that the paths of light from different directions—below, to the side, and above—from this distant star get bent just enough that they get focused so that, as a planet drifts past your field of view, the starlight greatly increases in brightness. It spikes, and then drops back down again. We know exactly what that rise, and what that fall, should look like. This is called gravitational microlensing. It's as though the planet's gravity served as a lens to focus the light of a distant star behind it.

One challenge for this effort is that it's a non-repeating thing. The planet just drifts by and keeps going on its way. Plus, the chances of a planet exactly crossing the view of a star, given the vast emptiness of space, are very small. You've got to monitor and track millions of stars, and some stars change their own brightness according to their own habits. You have to know what that change looks like, versus what this change looks like. It's called gravitational microlensing. We're building a catalog of how many planets are out there floating through space simply because we see this effect of the light from a distant star. That's another way to detect them, but again you're not waving to aliens in this effort. You're just sort of identifying the fact that there are planets out there, and you want to do a little better than that. We want to know how much mass does the planet have, how big is it, and could it look like Earth.

The most tried-and-true way that we have used to discover planets—and the way that's responsible for all of the hundred planets that are now known outside of our solar system—uses a principle first discovered by Christian Johann Doppler. He did this back in the nineteenth century. He was a German physicist and he did a very simple experiment. We've all noticed this and he noticed it with train whistles back when trains were more common than cars. He noticed when a train tweeted its whistle as it approached, the pitch of that whistle sounded different as the train approached than it did as it receded. The pitch changed. Either the conductor has a whole symphony of whistles and he changes them just as he goes by you, or there's something more fundamental going on—because no matter where you are you hear this. He's not doing this for your benefit, and trains really do only have one whistle or a few; but they don't have a symphony of them constantly changing frequency as they go by you. They have better things to worry about on the train.

The pitch goes—I can't imitate a train whistle so I'll imitate cars driving by swiftly at a racetrack. It's the same difference because the tires make sound on the track. Ready? neeeeeeEEYOWWwwwww. We know that sound. We've heard that sound before. Have you ever questioned it? Have you ever wondered, "Why does it sound different coming toward me than it does going away?" Maybe you hadn't because you just accept it as just what goes on in life. Hear it again. neeeeeeEEYOWWwwwww. It doesn't go neeeyneeey. It doesn't go yyyyeeeiiiiiii. It doesn't do that either. Doppler measured that first, in the 19th century. He didn't just simply assert that it was so; he found a mathematical form that described exactly the rate that that frequency changes, and it's a basic property of waves that happen to be sound waves emanating from an object in motion. It also works for light waves.

How do we invoke this for stars? You have a diagram and a telescope looking toward a host star. In the figure you notice two sort of yellow suns but that's not a double star system. That's one sun moving around a dot in space. What's going on there? Something I haven't told you, yet. If you're a star and you've got a planet in orbit around you, typically people think the star is in the center and the planets go around the star. No. It's not what's happening. What's happening is the star and the planet orbit their common center of gravity, which generally is not at the center of the star. If you can't see the orbiting planet because it's too dim, too far, too whatever, but you certainly see the light of the host star; you monitor it and you notice that it has these Doppler shifts. Sometimes the light is coming toward you, sometimes the object emitting the light is moving away from you—just like the train moving past.

You can measure the change in the frequency of the light, which is also a change in its wavelength. The two go together: high frequency, low wavelength. Here's the star doing this in reaction to the fact that a planet is in orbit around it. That movement of the host star has the same period as the period of the planet. The amount of that motion is a function of how much mass the planet has. The bigger the mass of the planet, the more the host star is going to move. It's a very effective means by which we can measure the existence of planets. Not only could you find one orbiting planet, you could find multiple planets this way; because, if you look at the variation of the host star you say, "Okay, I've got this 12-year period in there. There must be a planet at 12 years. Wait a minute. There's another level of jiggles on top of that 12-year jiggle." The host star is not only reacting to one planet, it's reacting to another planet combined. You have to tease one from the other, find the two signatures and then you get to report that some planets have multiple star systems.

This is making evening news. We've seen this Doppler effect, maybe not with light and not with train whistles, but with a police officer with the radar gun. In this case, he is using microwaves. I bet you didn't know microwaves were beaming you when you get your speeding ticket. It's a microwave gun using basically the same principle. He can measure your speed if you're coming toward him or if you're if you're moving away. It's just harder for him to catch up with you if you're receding, so he tries to get you as you're coming toward him—the old radar gun.

The first set of planets discovered by this method was Jupiter sized, massive. That's kind of to be expected. You're going to find the easiest data to acquire from a big tug from a big planet, not a tiny tug from an Earth-size planet. The first 50, 60, 70 planets were all Jupiter-sized planets in orbit around the host star, not surprising. It's like looking for your car keys under a lamppost. That's the easiest place to find them so you look there first. If they're not there you look somewhere else, but if you find them there you're doing good.

Here's another problem. I've got the list here for the first set of these planets that were discovered. Some of the first three were planets in orbit around the star 51 Pegasee, 47 Ursa Majoris, and 70 Virginis. These are code. Code for 51 Pegasee is the fifty-first brightest star in the constellation Pegasus. 47 Ursa Majoris is the forty-seventh brightest star in Ursa Major, a piece of which is better known as the Big Dipper. 70 Virginis is the seventieth brightest star in the constellation Virgo, Virgo the Virgin. These stars had Jupiter-sized planets, not orbiting where our Jupiter is, but really up close in a place that we never even imagined they could be.

We had to throw out all our theories of planet formation because we only had one solar system and we figured we must be representative. Everybody else must look like us so our theories must produce a solar system that resembles us: rocky planets in close, massive gaseous stars further out. None of them thus far look like our solar system. That's puzzling. Is it our method that prevents us from seeing? Maybe. We've only been doing this for about seven, eight, nine years. Massive planets orbiting much farther away from their host star, can take anywhere between eight to 20 years to go around their host star. We don't even have a baseline yet to measure a full orbit of solar systems that might have a very distant massive planet. The jury is still out, although it is still puzzling. It's got people wondering.

We've got hundreds of these across the sky. In an artist's rendition, there is a double planet star system just trying to imagine what it might look like. These are the times you need artists because you feed them the raw data and they show you what it might look like for your dreams and for the next generation of science fiction films. If you add all of these up there's a hundred of them across the sky. They're all over the sky. They're not just in one spot. They're everywhere, which is encouraging. Recognizing, of course, that planets are everywhere.

How do you get massive planets in close to the host star? This was a mystery for a while but we have some clever theorists who noticed that, if you're a massive planet and your solar system is still forming, there's a lot of debris between you and the host star. Depending on the orbital dynamics, you can actually exchange places with material. Fling material out as you migrate closer, and in the act you're sort of vacuuming out the interior of the solar system, exchanging angular momentum. The act of doing that gets you closer to the host star, evacuating the area in the process. This is part of the dynamics that enables planets to be flung to and fro, because the massive planets are wreaking havoc on the stability of the system.

Are we ever going to visit these planets? They're far, even the closest one. If we went 1000 miles per second—imagine traveling that fast. That's 100 times faster than any human has ever traveled, ever. Let's go at that speed—it would take you 800 years to reach the closest star to the Sun, and that one doesn't even have any planets. If you want to reach one that has planets, it would take you 2000 years. You'd be long dead before you got there. We still might want to make a catalog of these in case one day we figure out how to do it. One rule of space travel is you want to get to your destination before you die of old age—this is a criterion. You want to pick your planets carefully.

There is a new plan for NASA to build what we call a Terrestrial Planet Finder (TPF) and combine that with another one, SIM, the Space Interferometry Mission. These are missions up in space, which have a bank of telescope dishes working in harmony to give you the sharpest possible image of the smallest possible objects. We want to get an image of an Earth-like planet. We want to see the oceans. We want to see the textures. We want to measure the atmosphere. If we whip out a spectrograph, we can measure the chemistry of the atmosphere. Now you're talking, because the chemistry of the atmosphere is a product of what's going on, on the surface.

We have oxygen and we breathe it, but oxygen is a smoking gun of the fact that we have metabolism going on here on Earth. Without the metabolism, the oxygen would go away; it's unstable in anybody's atmosphere. Methane is unstable in the presence of oxygen. We've got methane around because it's got a constant source from farm animals, among others. Ozone is another signature.

If I have to pick a star to visit, I'll go first to one that has these signatures. That's where I'm going to go first.

Our spectrographs might also detect chlorofluorocarbons, ozone-destroying chemicals. It might find hydrocarbon contaminants, or soot from global deforestation. Suppose it finds that. That would be a sure sign that on that planet, there's not a single sign of intelligent life.

# Lecture Twelve
## The Search for Life in the Universe

**Scope:**

No question looms larger in the minds of the public than "Are we alone in the universe?" We don't know the answer, but we certainly have enough information to engage in a fertile discussion of the topic. In the early 1990s, when we started discovering planets beyond our solar system, the prospect of finding life dramatically increased and interest in this question was renewed. We close this series of lectures by examining the very real possibility that life exists elsewhere in the cosmos and speculating about its origins and chemical makeup.

## Outline

**I.** On Earth, life teems everywhere, but is it presumptuous to assume that such fertility exists elsewhere in the universe? Even if the ability to support life is rare, the universe is so vast that we have a huge area to sample and an extraordinary number of possibilities.

   **A.** As we said, the cosmos contains one sextillion stars, each of which could have planets. There are more stars in the universe than grains of sand on all the beaches on Earth.

   **B.** It would be egocentric to presume that we are alone in the cosmos, even if life is exceedingly rare.

**II.** The anthropic assumption that human beings are somehow special has misled people for generations and reversed the progress of science.

   **A.** In the early 1500s, Nicholas Copernicus, in his book *De Revolutionibus*, established the *Copernican principle*: that the Sun is in the center of the known universe, instead of the Earth. He asserted that Earth was no more important than the other planets in the solar system, nor was the Sun a particularly important star.

   **B.** Copernicus was not the first thinker to make this assertion. In the third century B.C., the Greek philosopher Aristarchus also speculated that the Sun was in the center, but his idea never caught on. The geocentric view of the universe made its way into the teachings of Aristotle and, later, the Catholic Church.

**C.** We now know that we're not even in the center of the Milky Way, nor is the Milky Way in the center of the universe, even though all other galaxies seem to be receding from us. We learned from Einstein that no matter where we look in the universe, galaxies will seem to be receding from that point.

**III.** What can we learn from the biodiversity of life on Earth?

**A.** The range of life forms on Earth is extraordinary. Think, for a moment, of the following random list of life forms: rhinovirus, algae, beetles, sponges, jellyfish, snakes, condors, and giant sequoia. Add to this list trilobites, which ruled the planet 500–600 million years ago, and dinosaurs. It's hard for us to believe that all these life forms come from the same universe, much less the same planet.

**B.** Imagine describing a snake as an alien life form: Many snakes stalk their prey with infrared detectors. They are capable of swallowing prey that is five times the size of their heads. They have no arms, legs, or other appendages, but they can slide along the ground at a rate of two feet per second. Such a creature is a good candidate for an alien, yet it exists here on Earth.

**C.** Hollywood is embarrassingly unimaginative in its portrayal of the diversity of aliens. Most of its creations have arms, legs, heads, fingers, and so on. These aliens are basically identical to humans, especially compared to the array of life forms listed above. Life on other planets should look as different compared to humans as other life forms on our planet look in comparison.

**D.** Note, however, that we have DNA in common with every life form on Earth.

**IV.** What do we know about the science, or chemistry, of life?

**A.** We know that life forms on Earth are based on organic chemistry, and a key ingredient of organic chemistry is carbon. We are carbon-based life.

**B.** Carbon is produced in vast quantities inside of stars and released into the galaxy, enriching clouds that then collapse and form stars, planets, and ultimately, people.

**C.** Chemically, carbon combines in more ways to make more kinds of molecules than all other molecules that exist. If you were to pick an element on which to base the diversity of life, carbon is the prime candidate.

**D.** Life, which is opportunistic, relies heavily on the fact that carbon can bond with itself and other elements in many ways, including ways that we are still discovering.

**E.** Carbon is so versatile and plentiful that we can state with a good deal of certainty that other life in the universe will probably be based on carbon chemistry.

**F.** Also abundant in the cosmos and in humans are hydrogen, nitrogen, and oxygen.

   **1.** We owe the abundance and distribution of these ingredients to the remnants of stars that have exploded, such as the Crab Nebula, a nebulosity formerly contained in a star that blew up 2000 years ago.

   **2.** The Crab Nebula is still expanding into space. Each of the color variations we see in images of this nebula represents a different chemical species manufactured in the thermonuclear crucible that was once the center of this star. Without these heavy elements, which are still traveling through the cosmos, the formation of planets with any ingredients other than those that were products of the Big Bang would be impossible.

**V.** How special is life on Earth?

**A.** We've all seen lists of the chemical ingredients of human life. We know that the human body is 80 percent water and, therefore, it contains more hydrogen atoms than anything else. Next in order comes oxygen, carbon, nitrogen, and so on.

**B.** The same list of ingredients for the universe would match one for one with those in the human body. First in the universe is hydrogen, followed by helium, which is inert; it has no chemical utility. These are followed by oxygen, carbon, and nitrogen…

C. The conclusion is that humans are *of* this universe. If we were made of extremely rare ingredients in the universe, we might have an argument for our uniqueness, but that is not the case.

VI. Where should we look for another planet that might have life?

A. Life as we know it requires liquid water. We don't know if that is a universal need.

B. Originally, we looked for planets that might sustain life in a certain habitable zone, not too close or too far from their host stars.

C. We have since had to expand the habitable zone, because of observations from Europa, a moon of Jupiter.

1. The surface of Europa appears to be frozen, but it experiences stresses from the tidal forces of Jupiter and its 20–30 satellites. Those stresses impart heat into the interior of Europa, which seems to be melting the ice in some places and causing shifting patterns in the ice on the surface.

2. We believe that Europa has an ocean that has been in existence for a billion years.

3. Life on Earth achieved major advances in our oceans. If we're looking for other life in the universe, we might need to start the search in our own backyard.

D. Mars shows evidence of the past existence of running water. We can see dried riverbeds and flooded plains, as well as evidence that some kind of material was carried in these floodwaters, but there is no sign of life on Mars now.

1. What happened to the water and atmosphere of Mars? We don't know. The water could be subterranean now, under the permafrost.

2. If Mars once had running water, it may once have had life. Any expedition to Mars should include a geologist and a paleontologist to search for fossils.

VII. Of course, the ultimate question about other life in the cosmos is, what is the chance that alien life is intelligent?

A. A few years ago, a rock was found in Antarctica thought to be from Mars and showing evidence of single-celled life.

1. Something big, such as an asteroid, must have hit Mars a long time ago, and pieces of its material were thrust out into space.
2. Perhaps Mars, with its wet history, formed life before life on Earth. If so, perhaps that single-celled life stowed away on a rock that landed here four billion years ago. If that's true, then all life on Earth has a Martian origin. This idea is called *panspermia.*
3. If life on Mars had DNA, we won't know whether DNA is an inevitable molecule of a chemical soup or whether we got our DNA from Martians in the first place.

**B.** One way to think about these questions is the *Drake equation,* attributable to American astrophysicist Frank Drake. This equation is not a typical mathematical formula, but a sequence of probabilities that affect the search for life.
1. First, we ask, what percentage of stars lives long enough for life to evolve?
2. Then we ask, of those stars, what percentage has planets?
3. Of those planets, what percentage is in the habitable zone?
4. Of those in the habitable zone, what percentage gave birth to life?
5. Of those that have life, what percentage evolved life to intelligence?
6. Of those that evolved intelligent life, what percentage developed technology that would enable civilizations to communicate across the galaxy?

**C.** To "solve" this equation, we set up the probabilities, and then multiply them out to arrive at the number of possible civilizations with which we might communicate.

**D.** We may need to modify some of the terms of this equation since its development. As we said earlier, the habitable zone seems to be broader than we originally imagined. Further, we know that life can thrive based on geochemistry rather than on energy from the Sun; this is revealed in the life forms that exist deep in the ocean, in boiling water heated by magma from the center of the Earth. The number of places we can find life and the hardiness of life are greater than what Drake imagined.

**VIII.** If we find other intelligent life, how will we communicate with it?

   **A.** We can't travel to other likely planets, because they are simply too far away, but we can send out radio waves, which travel at the speed of light. The language we use for these communications is mathematics.

   **B.** The duration of technology is a factor in communication in the cosmos. Suppose alien signals arrived on Earth 100 years ago. The aliens would assume that no intelligent life existed here, because we had no way to respond to their signals. We have been technologically capable of responding for only the last 50–75 years out of 4.5 billion.

   **C.** We have been communicating with the galaxy unwittingly for the past 60 years by sending out television signals. Alien anthropologists may now be decoding the *Howdy Doody Show*, *Amos and Andy*, and *I Love Lucy* as the earliest signs of intelligence on Earth.

**IX.** As we conclude these lectures, I hope "my favorite universe" has now become "your favorite universe." I will leave you with this thought: "In life and in the universe, it is always best to keep looking up."

**Suggested Reading:**

Goldsmith, Donald, and Tobias Owen. *Search for Life in the Universe*. Mill Valley, CA: University Science Books, 2001.

Margulis, Lynn, and Dorian Saga. *What Is Life?* New York: Simon & Schuster, 1995.

McKay, D. S., et al. *Search for Past Life on Mars*. Washington, D.C.: American Association for the Advancement of Science, 1996.

**Questions to Consider:**

1.  In what fundamental way do Hollywood aliens show a deep lack of imagination?

2.  Silicon and carbon are chemically similar, yet life as we know it is based on carbon. Describe two reasons why carbon is so useful to life and why silicon-based life may be much rarer than carbon-based life.

# Lecture Twelve—Transcript
## The Search for Life in the Universe

Welcome back to this, the final installment of My Favorite Universe. I've saved for last, the subject that occupies a disproportionate fraction of my intellectual energy, and I know from experience—my empirical encounters with people of the public—that it also occupies your mind as well. That subject is, of course, the search for life in the universe. No question looms greater in the minds of the public than that question, "Are we alone?"

We don't know the answer to that question yet, but we have enough information to make for a formal discussion about how to think about that question and how to think about the answers. This all began with a renewed fervor just about 10 years ago, the early 1990s that was, when we started discovering planets beyond the planets of our solar system. Clearly, if you're going to imagine life somewhere other than Earth, you imagine it on a planet, not somehow living inside a star. The discovery of planets fueled that curiosity and that act of having found planets—now that number rising through 100; more planets known outside of our solar system than within it—the prospects of finding life dramatically increased, just by the fact of this knowledge.

Generally, it's bad practice to make sweeping generalizations because life on Earth is our only example of life anywhere. To presume just because of what we know on Earth—and Earth is teeny, and no matter where you look on Earth, no matter under what rock you turn, there's life there—to presume that such fertile circumstances exist elsewhere in the universe may be a little strong until you realize how big the universe is. Even if life were something rare, all you need is a big enough sample of occurrences. All you need are enough stars and enough planets that, no matter how rare this thing is we call life was on earth, you figure it's likely to have formed some place else. These numbers? The Sun has a bunch of planets of its own: eight, nine, 10 planets. It's got a lot of planets. The Sun is an ordinary star. There are 100 billion stars in our galaxy, 100 billion galaxies in the universe—a sextillion stars. There are more stars in the universe than grains of sand on all beaches on earth. No matter how rare we want to think life is, we'd be inexcusably bigheaded, inexcusably egocentric to presume that we are alone in the cosmos.

These sort of anthropic assumptions have misled people for generations. Anthropic assumptions that somehow we are special, that we are privileged, that we have the centralized view on the cosmos. That has reversed the progress of science every time somebody comes up with such an idea. Right now we're a little more mature in this line of investigation, and what we do is invoke the *Copernican principle*. The Copernican principal was named after a famous, important poet/astronomer, Nicolaus Copernicus. He was born in the 1400s and did his greatest work in the early 1500s, culminating in the publication of a book, *De Revolutionibus*. What that book did was put the Sun back in the center of the known universe. Until then, Earth had been in the center. And there it had been in the hearts and minds of people ever since the third century B.C., when an early suggestion by the Greek philosopher Aristarchus was put forth that perhaps the Sun was in the middle of the known universe. It never caught on because it assumed, if the Sun was in the middle, then we are moving around the Sun, and if we're moving around the Sun we should see stars swinging back and forth because of a change in angle of view on them. Nobody saw that, plus it didn't feel like you were moving, so why should we believe the guy.

The geocentric view caught on, worked its way into Aristotle's teachings and Ptolemy's teachings and the teachings of the Catholic Church where it became codified as law. It was generally accepted and, of course, it was also basically self-evident. If you looked around it looked like everything was going around the Earth. It was not only self-evident, but surely God would have it no other way. We being Earth dwellers, and being special creatures in the mind of God, surely we occupied a special place in the cosmos.

What we came to learn by the work of Copernicus was that no, that's not how the world is assembled. The Sun is in the middle, and in fact, we're just another planet. We have no greater rank than other planets out there in the solar system. We're not special. The Sun is not special. We just happen to be really close to it, so it's important to us; but in the big picture, it's not special. We on Earth, as a planet, are not special and our star is not special. For a while, we looked around and thought we were in the center of the Milky Way galaxy. We must be. Upon acquiring better data we found out no, we're not in the center, we're not even on the edge. We're somewhere in the nondescript suburbs. If you look around, it looks like we're in the center of the universe. All the galaxies are receding from us in every direction; it looks like we're in the center. No, no. Let's not fall for that again. We've learned from Einstein's general theory of relativity that no matter where you are, it will look like you're in the center. It's a product of the fact that you have this enormous universe stretching in every direction and no matter where you are, it will look like you're in the center and everything is moving away from you. Like raisins in a raisin cake, the cake grows and whichever raisin you're on, you think you're stationary and all the other raisins are running away from you.

I think it's a conservative posture to say that Earth, and life on it, is not immune to the Copernican principle, just going on the basis of history. Let's use that as a first step.

What can we learn from biodiversity? On this one planet we call Earth there are extraordinary ranges of life forms. You can make the list forever, of course, but let me handpick a few just for the sake of this conversation. I want to handpick a few, and I'll tell you why I picked these in just a moment: the rhinovirus (the virus that gives you the common cold), algae, beetles, sponges, jellyfish, snakes, condors, and giant sequoias. I handpicked those from the animal and plant kingdoms. Line them up in size place. You have the virus over here, the giant sequoia over there. Check them out. Line them up.

When you look at these creatures—a jellyfish, a tree, and a snake— you would wonder whether they came from the same universe, much less the same planet. Look at that! Add to these creatures—that are all living creatures in modern times—add trilobites. Here we go. Here are fossil trilobites. There was a day when they ruled the world 500–600 million years ago. Put those on the scale, too. They're extinct, but put them there. That's life that used to exist here on Earth. We don't have to stop at trilobites. Come somewhat later and you get good old *Tyrannosaurus rex*, dinosaurs co-existing with this range of life. It's extraordinary.

Suppose you visited some alien planet and you came back giving the following account of some life form that you just saw on this planet. You come back to your friend and say, "You've got to believe me. I saw this animal. It can stalk its prey with infrared detectors and it swallows whole, live animals five times bigger than its head. It has no arms and legs or any other appendage, yet it can slide along the ground at about two feet per second." You say, wow, what an exotic creature that is. Well of course, you've just described a snake. Snakes have no arms or legs, not all but many have infrared detectors, and they can eat things five times bigger than their head. That's a scary creature if you haven't seen any other kind of creature in your life.

I'm thinking to myself, with this depth of biodiversity—if I may take just a quick diversion, this being the last lecture I'll give myself a diversion here—when I think of Hollywood aliens, how good a job are they doing? You know, they're embarrassingly poor in the diversity of form that they grant their aliens. There are some good examples; not all of Hollywood was bad. Take the 1950s classic, *The Blob*; that was a good one, *The Blob*. That didn't have any arms or legs but it was clearly alive, it was intelligent, it was a creature. That was good. Also, the film 2001 *A Space Odyssey*. They didn't even show the aliens. They just showed this monolith thing that they aliens used to communicate with. They didn't even go there and I thought that was brave, because everyone wants to see some lizard thing coming out of the mountain.

What do the rest of the Hollywood aliens have; they've got arms, they've got legs. They've got a head, shoulders, knees, toes, and fingers—maybe three fingers—but they've got fingers. They've got a neck, maybe a little longer. They've got a head, probably a bulbous forehead because we assume they're smarter than us. In the end, if these are your anatomical properties, you're basically identical to a human being compared with this list of animals I put up here in front of you just a moment ago. Every alien I've ever seen in a movie is identical to humans as far as the diversity of life on earth is concerned. I'm not that impressed.

I'm thinking, if we're going to find life on another planet, it better at least look as different compared to us as other life forms on earth look compared to us. It's got to look at least that different for growing up on another planet. We have DNA in common with every life form on this size scale. Jellyfish? Viruses? Sequoia? There's some percent of our DNA that is identical to all the life that's on this form. If we go to another planet and look at their DNA or whatever is their magic chemical, I'm going to assume it looks more different. I think I have the right to assume that, but that's just the diversity of life.

There are certain things about the science of life that I think are inescapable, and that would be the chemistry. By definition, we are made of organic molecules. A key ingredient in organic chemistry is carbon. We are carbon-based life. Carbon, you may remember from an earlier lecture, is produced in vast quantities inside of stars, thrust out of the stars by mechanisms that release the outer layers of a star or just blow up to release into the galaxy enriching clouds that then collapse and form stars and planets and people and everything in between.

Carbon chemically combines in more ways to make more kinds of molecules than all other molecules that exist. In other words, if you were to pick an element upon which to base the diversity of life, there is no contest; carbon is your element. Life, which is opportunistic, relies heavily on the fact that carbon can bond with itself and with other elements in many and varied ways, ways that we are still discovering. There's the famous Buckminster Fuller ring, the big molecule of carbon with 60 carbon atoms making a sphere that looks like a soccer ball. You can take that and make tubes, make Bucky tubes out of it. Diamond is carbon. The lead in a lead pencil is carbon. Carbon is a versatile element; it's also plentiful in the cosmos. If we're going to find life based on some kind of chemistry, if I'm a betting man; I would tell you it's going to be based on carbon chemistry. I don't know what it will look like, three eyeballs, two heads, antennae, but it's going to be based on carbon chemistry. The sheer fertility of it is inescapable.

There's not just carbon out there. We've got hydrogen in us, which we owe to the Big Bang. The Big Bang endowed the cosmos. Ninety percent of all atoms are hydrogen atoms. That's convenient. We also have nitrogen and oxygen. These are abundant elements in the cosmos. In fact, we owe the distribution of these ingredients and the existence of the ingredients in the first place to the remnants of stars that have exploded. We have the Crab Nebula, a nebulosity formerly contained inside the core of a star, a star that blew up. This one was observed to blow up 1000 years ago, and that's what remains of it today.

It's still rapidly expanding into space and each of those color variations in the image represent a different chemical species manufactured in the thermonuclear crucible that was the center of that star before it exploded and during the explosion. It spread out into the galaxy, enriching the next generation of the formation of stars. Allowing solar systems to have more ambitious things than just a star in their center. Without these heavy elements, you couldn't make planets with surfaces that contain ingredients that were not products of the Big Bang itself.

Here's something interesting. How special are we? You still want to think we're special, don't you? Check this out. Let's list the ingredients within the human being in order of frequency. The number one element in the human body—the number one molecule is water; we knew this from biology class. Some 80 percent of your body mass is water. Fine. What is water made of, hydrogen: $H_2O$. We have more hydrogen atoms in our body than any other kind of atom.

What's next in our body; oxygen is next. Carbon is third, next is nitrogen and you go on down the list. I bet you didn't know that if you made the same list for the entire universe, it would match one-for-one the ingredients of the human body and all life on Earth. Hydrogen is the number one ingredient in the universe. Number two is helium. Helium is chemically inert. You can't do anything with it anyway, except inhale it and sound like Mickey Mouse or something. It's fun to play with helium but it has no chemical utility. Next in the universe is oxygen. After oxygen is carbon and then nitrogen right on down the list.

We are of this universe. If we were made of bismuth or some isotope of uranium—extremely rare ingredients in the cosmos—you'd have an argument. You'd say we are special because you don't find these ingredients just any place. You only find it here on Earth; but that's not the case.

Where would you go to look for a planet that might have life? Life as we know it requires liquid water. If you're going to look around for a planet, you want a planet that's not too close to its host star, because whatever liquid water it might have had would have evaporated. If it's too far away, it would freeze. You want it to be just right. It's the Goldilocks effect. You want it to be in a sort of habitable zone around a host star. In that way you have a chance of finding life that resembles us in some fundamental way. Surely it will resemble us in chemistry, but I don't know if the need for liquid water is a universal truth or not. We know we need it and we exist, so why not impose that, at least, in the first pass in our search for life.

This notion of a green zone, a habitable zone around a host star, not too close and not too far, is a little naïve, it turns out. We've learned that Jupiter's moon, Europa—Jupiter is way outside of the habitable zone—Jupiter's moon, which is covered in ice, is actually being stressed by the tidal forces of Jupiter and the surrounding satellites. Jupiter has 20–30 satellites. That stressing of the physical body of the moon Europa imparts heat into the interiors of this moon, and that heat melts the ice. If you look from one week to the next, one month to the next at the frozen ice patterns on the surface, the next week it has shifted as though it's being carried on a liquid under layer. We think there's an ocean on Europa; we think it's been there for a billion years.

Good evidence shows that life had some major advancing that it did within our own oceans. If we're going to look for life in the universe, let us start in our backyard. I want to go ice fishing on Europa. I want to melt down through the surface and put a little submersible and look around and see if anything swims up and licks the lens. That's what I want to do.

Let's look, not only at Europa, but also Mars. Some experiment went real bad on Mars. Mars has all kinds of evidence of running water, dried riverbeds and flooded plains where you see that material had been carried down on flood waters. There's not a sign of life there now. Nowhere on Martian surface do we see life. Where did the water go? We think it's subterranean, a permafrost. Where did the atmosphere go that once supported the water? Nobody knows. Some knob got turned, rendering Mars wholly inhospitable to life. It's the same with Venus. Venus is 900 degrees Fahrenheit. The planet has the same mass and the same size as Earth. Some experiment went wrong, a runaway greenhouse effect. What are we doing on Earth that might have Earth end up like these two planets as we go forth? We'd better keep an eye on that.

In any case, if Mars once had running water, maybe it once had life. If it once had life, if I'm going to go to Mars, I'm going to bring a geologist and a paleontologist with me, because I don't know how to go rummaging through rocks and find fossils. I know how to look up; I've never spent much time looking down. I'm going to bring them with me to check for fossil evidence.

How about intelligence? What is intelligence? At the end of all of our questions of is there life, yes or no, there is the question: If there is life, is it intelligent? That's an important question. What is intelligence? We have a hard enough time defining it in ourselves and in other species to suggest that we would know it and be able to define it for an alien.

Getting back to Hollywood just for a brief moment. I have seen some stupid aliens in the media. I remember I was driving from New York to Boston and I had on AM radio and there was a radio play. It was fun listening to a radio play and thinking of what my parents did before TV. Here I am driving and aliens were swooping down on Earth. Why? They were sucking up the water. Why? Because these aliens thrived on hydrogen and they needed the hydrogen in the $H_2O$ of Earth's oceans, and this was creating global catastrophe. They had to fight off the aliens, and I'm thinking these are stupid aliens. Ninety percent of all atoms in the universe are hydrogen. You don't have to come to this little pipsqueak planet and suck up the thin layer of $H_2O$ that resides on its surface. How about the thousand solar-mass gas clouds you drove past on your way to Earth? Why don't you wake up and take an introductory astronomy class or something before you start running around sucking up Earth water.

I've got another one. Do you remember *Close Encounters of the Third Kind*? It was a brilliant film, nicely timed and everything, but I've got a pet peeve. There's a scene where the aliens are announcing where they're going to land, but nobody knows this. They're sending signals to their Teletype, back before they had computer monitors. The Teletype is typing out this series of numbers and they say, "It's been sending this signal and we don't know what it is." And some clever guy says, "I know what that is. That's the longitude and latitude on Earth. I bet that's where they're going to land." They went to Devil's Tower, Wyoming or wherever the place was, and set up a runway and lights; and sure enough, they landed there.

I'm thinking, how do they know longitude and latitude of Earth? Latitude goes from zero degrees to 90 degrees. That's kind of arbitrary; it could have gone to 100 degrees. That's just our own invented line. Longitude is especially arbitrary. That was a political decision from 150 years ago. The Prime Meridian goes through London and it goes through Greenwich in England. It might have gone through Paris. Hell, it could have gone through Hawaii, if

Hawaii had kept records at the time. It's completely arbitrary. To know what our longitude system is means you know our culture so well you know all of the politics of what led to it in the first place. If you know that much about our politics, why not just send over the Teletype, we're going to land to the left of Devil's Tower Monument. Send it in English, for goodness sake, if you know that much about our culture; but they're got to do the mysterious way. The way is just stupid; if you know that much just give us some other kind of sign instead of cryptic numbers in base 10. Come on now. Plus they had a runway. If you come from another galaxy in a flying saucer, you don't need a runway. Runways are for people who use the air for lift. Flying saucers don't use the air for lift; that's why they're flying saucers. We need more people in Hollywood to take this video course.

How does life get around? We've got life here on Earth. You may remember the news stories from a few years ago. We found a rock on Earth that came from Mars, and it had tantalizing evidence of there possibly having been life inside the cracks of the rock, stowaways that journeyed through the void and vacuum of interstellar space, landed on Earth, landed in Antarctica and was collected in the Allan Hills, a place where people go meteorite hunting. Meteorites fall all over, but if it's in the ocean you're not going to find it. If it's in a forest you're not going to find it. It will look like any other rock. If it falls on an ice sheet, it's a meteorite. Ice sheets don't have rocks sitting on them, generally.

How did it get there? Something thrust it off of the Martian surface. That's curious; because if that's possible, it meant that Mars was hit by something really big a long time ago and pieces of its material were thrust into space, floated and landed here on Earth. That tells me that maybe Mars, with its wet history, formed life before life formed here on Earth. If it did, maybe that life stowed away—you just need single-cell life—on a rock that landed on Earth four billion years ago. If that were the case, it would mean that all life on Earth has a Martian origin, making all of us Martian descendents. That's entirely possible. It's called *panspermia*. If we go to Mars and find life and find that the life there has DNA, we won't know whether DNA is an inevitable molecule of a chemical soup or whether we got our DNA from them in the first place.

One way to assess how to think about these questions is the *Drake equation*, attributable to American astrophysicist Frank Drake. He came up with an equation that enables you to organize your ignorance. It's not an equation in the classical sense of a mathematical equation; it's just a sequence of probabilities that affect the search for life. You ask, what percentage of stars live long enough for life to evolve? Of those stars, what percent have planets? Of those that have planets, what percent have a planet in the nice zone? What percent of those in the nice zone actually gave birth to life? Of those that had life, what percent evolved life to intelligence? Of those that had intelligence, what percent evolved a technology— because you're not talking across space without some kind of device? You're not sending smoke signals across the galaxy; you need something a little more sophisticated.

You can set up these probabilities, multiply them all out, and see, starting with the total number of stars in the galaxy, how many civilizations might you communicate with.

Some of the terms of that equation need to be modified. This green zone is now broader than we had ever imagined. If you could warm up a moon of Jupiter through means that have nothing to do with the Sun, your source of energy is quite fertile. Your sources of energy are many more in number than simply the Sun. That is so, "on Earth." There are bugs and life forms that thrive in boiling liquids in the undersea vents, hot water that is deep down, heated back where you have magma. It is heated, boils and gurgles up. Dissolved salts are there and it deposits those salts upon hitting the cold bottom of the ocean into these chimneys, and in those porous chimneys are life forms, thriving on geochemistry rather than on the Sun. The number of places we can find life, and the hardiness of life, is far greater than anything Frank Drake imagined in the first place.

How are we going to talk to them? It's too far to travel, based on any way we know how to get there. It would take you 75–100,000 years just to get to the nearest star. Forget it. We can use radio waves moving at the speed of light. What kind of signal are you going to send? Do you say, "Hey, I'm here?" Do you say, "*parle vous francais?*" No, no. Of course they don't speak our language, not our normal language derived out of our national borders, no. If they speak any language they're going to speak math. Math is the language of the cosmos. We have to be clever about how we construct some little math statements just to compare to see if we're speaking literally and figuratively, the same language.

The duration of technology matters. Suppose aliens try to send signals to us and the signals got here 100 years ago. A hundred years ago we didn't have technology to receive and decode and send back signals. They would presume there was not intelligent life on Earth or at least not intelligent life that had technology, because we'd have a way to respond. We've only been technologically capable for about the last 50 or 75 years out of how long—four and a half billion! That's nothing. Make a timeline of that and throw a dart. Most of the time if that dart hits the Earth timeline, Earth will just have boring, single-celled creatures trying to change the atmosphere of Earth from its carbon dioxide ingredients to oxygen. That's what's going to happen most of the times you throw the dart. Maybe sometimes you'll hit the last 100,000 years; it's still a pretty uninteresting time technologically. We're running around buck naked through the forests. There is a time after the oxidizing atmosphere when life gets kind of interesting. In the last 30,000 years of this 4.5 billion-year timeline you'll catch cavemen drawing pictures. You're not going to exchange technology with them. You might compare and contrast paintbrushes, but that's about it.

Actually, we have been communicating with civilizations unwittingly. For the past 60 years our TV signals have been emanating from Earth, moving at the speed of light. AM radio doesn't escape the Earth, the ionosphere reflects it, but FM radio escapes. Is anybody listening? There are a few planets that we have discovered in that zone. What would they be hearing? Alien anthropologists would be decoding the earliest signals that came from Earth, the earliest signs of our intelligence. They'd be decoding the *Howdy Doody Show*. They'd be decoding *Amos and Andy* and *I Love Lucy*. This is how they would learn how humans interact with each other. This is what they would surely use to deduce that there are no signs of intelligent life in the cosmos.

So ends 12 lectures of "My Favorite Universe," each drawn from one of my monthly essays that appeared in *Natural History* magazine. My goal, however, was to turn the subject of this series into "your favorite universe." Consider these lectures a kind of hors d'oeuvre, perhaps piquing your interest in the cosmos just enough to have learned that in life and in the universe, it's always best to keep looking up.

# Timeline

**Age**

0 (13.5 billion years ago) ............... Big Bang; the space, time, matter, and energy of our universe burst forth into existence from a small, dense, hot fireball.

Before $10^{-43}$ seconds ............... Planck era; quantum foam prevails; space and time are entangled; all forces of nature are merged.

$10^{-43}$ seconds ............... Gravity separates and becomes distinguishable from other forces of nature.

$10^{-36}$ seconds ............... Strong nuclear force becomes distinguishable from electroweak forces, triggering a period of rapid inflation.

$10^{-32}$ seconds ............... Inflationary era ends; era of quark formation begins.

$10^{-12}$ seconds ............... Split of electromagnetic and weak forces, leaving the four familiar forces of nature.

$10^{-6}$ seconds ............... Quark-hadron transition, where quarks combine to form protons, neutrons, and other baryons.

0.02 seconds ............... Formation of electrons and positrons.

3 minutes ............... Expanding, cooling universe "freezes" out 86 percent protons and 14 perecent neutrons.

35 minutes ............... Less than 1 in $10^8$ excess of electrons over positrons leaves a universe in which matter dominates over antimatter.

380,000 years ............... Remaining free electrons combine with nuclei. Universe becomes transparent to light. Distribution of matter at this moment leaves a signature in the pattern of photons, measured as the cosmic microwave background.

20 million years ..........................Era of the first stars in the universe.

3 billion years ..........................Birth of Milky Way galaxy.

8 billion years
(4.6 billion years ago).......................... Birth of solar system. Period of heavy bombardment begins, raining leftover debris from interplanetary space on the existing planets.

4.0 billion years ago ...................Period of heavy bombardment ends.

5000 years ago .........................Stonehenge built; as a monument to the cycle of the seasons, Stonehenge is regarded as one of the first astronomical observatories.

**Date**

384–322 B.C. ...............................Aristotle, Greek philosopher; advanced the idea of the geocentric (Earth-centered) universe.

310–230 B.C. ...............................Aristarchus of Samos, Greek astronomer and mathematician; first proposal of a heliocentric (Sun-centered) universe.

c. 85–165 C.E. ............................Claudius Ptolemy, Greek astronomer and geographer; perfected the geocentric universe with a full system of epicycles depicting the cyclic movement of all astronomical bodies around the Earth.

476 C.E. .....................................Fall of the Roman Empire.

| | |
|---|---|
| 500–1000 | Dark Ages in Europe. Plague ravages population; witches burned; heretics disemboweled. |
| 900–1100 | Baghdad is the world's center of learning and scholarship in astronomy, mathematics, and medicine. |
| 1473–1543 | Nicolaus Copernicus, Polish astronomer; on his deathbed, he publishes *De Revolutionibus*, containing a fertile prescription for a heliocentric universe. |
| 1546–1601 | Tycho Brahe, Danish astronomer; measured planetary positions with unprecedented accuracy from his observatory. |
| 1564–1642 | Galileo Galilei, Italian physicist and astronomer; father of modern science in methods, tools, and philosophies. |
| 1571–1630 | Johannes Kepler, German mathematician. |
| 1600 | Giordano Bruno, Italian monk and astronomer, burned alive, naked and upside down, by the Catholic Church, for suggesting that Earth was not alone in the universe as a habitat for life. |
| 1608 | Hans Lippershey, Dutch spectacle maker, invents telescope. |
| 1609 | Galileo makes his first telescope. A year later, publishes *Sidereus Nuncius* (*The Starry Messenger*), reporting on his observations of the heavens, the first ever with a telescope, containing proof that Earth is not the center of all motion. |

1609 ........................................Kepler publishes his first two laws of planetary motion based on the data acquired and willed to him by Tycho Brahe.

1619 ........................................Kepler publishes his third law of planetary motion, the first predictive mathematical statements about the behavior of the universe.

1632 ........................................Galileo publishes *Dialogue Concerning the Two Chief World Systems*, a staged debate between the heliocentric and geocentric theories, mocking believers of the geocentric system.

1633 ........................................Galileo faces the Holy Roman Inquisition, is forced to recant, and is placed under house arrest.

1643–1727 ...............................Isaac Newton, English physicist; most brilliant and influential scientist ever to have lived.

1687 ........................................Newton publishes *Principia Mathematica*, containing the universal laws of motion and gravity.

1704 ........................................Newton publishes *Optics*, containing laws of optics that demonstrate that white light is composed of colors and advancing the idea that light travels in particles rather than waves.

1776 ........................................American Revolution.

1789 ........................................French Revolution.

1795 ........................................France advances and officially adopts the metric system.

1799–1805 ................................Marquis de La Place, French mathematician, publishes the five-volume *Méchanique Céleste*, extending the power of Newton's theory of gravity.

1855–1856 ................................James Clerk Maxwell, English physicist, advances his famous equations that describe the complete behavior of electromagnetic energy, creating a foundation for Einstein's relativity theories.

1895 .........................................Joseph John Thompson, English physicist, discovers electron.

1900 .........................................Max Planck, German physicist, introduces the *quantum*, ushering in the era of modern physics.

1905 .........................................Albert Einstein, German physicist, publishes "On the Electrodynamics of Moving Bodies," commonly known as his special theory of relativity, redefining our understanding of the relationship between space and time.

1916 .........................................Einstein publishes his general theory of relativity, a modern theory of gravity that supplants Newton's universal laws of gravitation. Describes a universe in which space tells matter how to move, and matter tells space how to curve. Allows for the existence of black holes and the Big Bang.

1918 .........................................World War I ends.

1918 .........................................Ernest Rutherford, New Zealand physicist, discovers the proton.

1919 .........................................Arthur Stanley Eddington, English astrophysicist, measures the bending of starlight around the Sun, confirming a basic prediction of Einstein's general theory of relativity.

1920s ..........................................Quantum mechanics developed by many physicists in America and several European countries.

1923 ..........................................Edwin Hubble, American astronomer, discovers distance to Andromeda nebula. It is a galaxy external to the Milky Way.

1929 ..........................................Hubble discovers expanding universe.

1932 ..........................................James Chadwick, English physicist, discovers neutron, completing our basic picture of the atomic nucleus.

1933 ..........................................Carl David Anderson, American physicist, discovers the positron, the first antimatter particle and the favorite fuel of science fiction writers.

1945 ..........................................World War II ends. America detonates three atomic fission bombs, one as a test and two in warfare against Japan.

1957 ..........................................American astrophysicists Margaret E. Burbidge, Geoffrey R. Burbidge, and William A. Fowler and English astrophysicist Fred Hoyle publish the landmark paper "Synthesis of the Elements in Stars," describing in detail, for the first time, the

Formation of heavy elements inside of stars via thermonuclear fusion. Calculations made possible by declassified military documents on nuclear energy.

1957 ............................................Soviet Union launches *Sputnik I*, the first artificial satellite. It is thrust into orbit based on the principles of gravitation first advanced by Newton.

1961 ............................................Yuri Gagarin, Soviet cosmonaut, becomes the first human to orbit around Earth.

1963 ............................................Martin Schmidt, American astrophysicist, discovers quasars, galaxies whose prodigious energy output is driven by supermassive black holes lurking in their cores and dining upon matter that comes too close.

1965 ............................................Arno Penzias and Robert Wilson, American physicists, discover the cosmic microwave background, formed 380,000 years after the Big Bang and permeating the expanding universe as a bath of microwave light.

1966 ............................................NASA's *Gemini* program releases an image of Earth from space.

1968 ............................................NASA's *Apollo* program releases an image of "Earth Rise" over the lunar landscape.

1969 ............................................Neil Armstrong and Buzz Aldrin, American astronauts, are the first to walk on the Moon.

1976 .............................................NASA sends two unmanned *Viking* spacecraft to visit Mars. They look for life but find none.

1990 .............................................NASA launches the Hubble Space Telescope, the first of the great spaceborne observatories, giving unprecedented clarity of cosmic imagery.

1994 .............................................Comet Shoemaker Levy-9 collides with Jupiter, acting as a shot across our bow and reminding Earthlings of the hazards of asteroid and comet impacts.

1995 .............................................Michel Mayor and Didier Queloz, Swiss astrophysicists, discover the first planet in orbit around a star other than the Sun.

1996 .............................................NASA releases the "Hubble Deep Field," the deepest image of the universe ever taken, showing countless galaxies with remarkable detail.

1996 .............................................Martian Meteorite ALH84001 shows possible evidence for life on Mars.

1999 .............................................American astronomer Brian Schmidt and others use light from distant supernovae to discover that the expansion of the universe is accelerating; this discovery implies the existence of a dark energy supplying a form of antigravity in the vacuum of space.

2002 .............................................More than 100 planets are now known in orbit around stars other than the Sun.

2003 ..........................................NASA's Wilkinson Microwave Anisotropy Probe measures the cosmic microwave background with unprecedented accuracy and precision, confirming the basic tenets of Big Bang cosmology.

# Glossary

**Accretion disk**: A swirling disk of gas that funnels down toward a neutron star or black hole and is drawn from a nearby star or clouds in interstellar medium.

**Air resistance**: The resistance and subsequent slowing experienced by a moving object as its surface encounters air molecules.

**Antimatter**: The complete opposite of regular matter. Antimatter has the reversed characteristics of regular matter; for instance, the electrical charge and spin of regular matter is completely reversed in an antimatter particle.

**Asteroid belt**: The region between Mars and Jupiter that is littered with chunks of rock and iron debris left over from the formation of our solar system.

**Asteroids**: Sometimes called minor planets, asteroids are chunks of rock and iron left over from the formation of the solar system. Some are the remains of shattered mini-planets. Although asteroids can be found on almost any orbit around the Sun, most orbit in the gap between Mars and Jupiter, an area known as the *asteroid belt.*

**Astrophysics**: The branch of science that seeks to apply the laws of physics to explain the past, present, and future of the universe and all its content.

**Atmosphere**: The gaseous envelope surrounding a planet.

**Atmospheric pressure**: The weight of a column of air that is the height of the atmosphere. Expressed as the ratio of force and area, as in "pounds per square inch."

**Atom**: The smallest part of a chemical element that retains the identity of the element. It is normally composed of electrons, protons, and neutrons.

**Aurora**: Curtains of lights in the upper atmosphere created by charged particles from the Sun interacting with Earth's magnetic field and molecules in Earth's atmosphere. Earth's aurorae are commonly known as the Northern and Southern Lights. Scientists have witnessed auroras on other planets in our solar system, such as Jupiter and Saturn.

**Big Bang**: A theory for the origin of the universe that has achieved broad support from experiments in nuclear physics and observations in astrophysics. Its basic premise is that the universe began in a small, dense, hot state followed by an explosion that brought space and matter into existence approximately 13 billion years ago. The universe, today, still expands from this explosion.

**Binary stars**: Two stars that form together and revolve around a common center of mass. As many as 50 percent of all star systems may be binary star systems.

**Black holes**: Regions of space and time where the gravity is so high that the fabric of space-time itself has warped back on itself, preventing escape by anything that falls in or tries to get out. The escape velocity of a black hole is effectively greater than the speed of light.

**Blue shift**: The shortened wavelength of light due to the motion of the light-emitting object toward you. Because motion is relative, this shift also occurs if you are in motion toward the object.

**Celestial**: Related to the starry sky as seen from Earth.

**Centrifugal force**: The outward force that an object feels when it revolves around any other object or position. It may also be considered the force that creates the tendency to "fly" off at a tangent. A centrifugal force is not a true force at all. It is just the tendency for the revolving object to move in a straight line—which is what the object would do if no force (like gravity or a tether) were acting to keep it revolving.

**Centripetal force**: The force that keeps an object revolving around any other object or position. The Sun's gravity provides the centripetal force that keeps all planets in orbit. Otherwise, they would fly away into interstellar space.

**Chemical bonds**: Bonds that enable atoms, by way of their outer electrons, to combine to form molecules.

**Comet**: A comet is often referred to as a "dirty snowball." Comets are made of mostly ice and dust and orbit the Sun primarily in a flattened region of the solar system beyond the orbit of Neptune, known as the Kuiper Belt, and a spherical region that extends halfway to the nearest stars, called the Oort Cloud. A comet's "tail" is formed when the icy comet, approaching the Sun, evaporates its outermost layers and encounters the solar wind blowing the gasses into an extended stream away from the Sun, no matter what the comet's trajectory.

**Complex molecules**: Ensembles of atoms, bound together into large molecules, such as proteins and nucleic acids, the building blocks of life.

**Constellation**: The random patterns of stars in space as seen from the Earth. The celestial sphere is segmented into 88 constellations. Each constellation has a name that, in rare cases, actually resembles the assigned star pattern.

**Corona**: The thin and vacuous outer atmosphere of the Sun, with a temperature of millions of degrees. Because it is considerably dimmer than the visual surface of the Sun, the Sun's corona can be detected only with specially designed telescopes called *coronagraphs* or during a total solar eclipse.

**Cosmic microwave background**: An omni-directional bath of microwave energy with a temperature of a few degrees Kelvin. The cosmic microwave background is a residual signal from the early formation of the universe, when free but obscuring electrons combined with atomic nuclei, allowing light to travel freely. This signal lends strong support to the theory that the Big Bang is responsible for the expanding universe.

**Cosmic object**: An object that resides in the cosmos or universe.

**Cretaceous-Tertiary boundary**: The geologic boundary between the Cretaceous and Tertiary periods in the fossil record of Earth history, dated to 65 million years ago and commonly called the *K-T boundary*. This important moment in time is characterized by an abrupt extinction of land animals and is believed to be the result of the impact of a large meteorite. This most famous of Earth's extinction periods left all classical dinosaurs extinct.

**Cubic inch, cubic centimeter, cubic foot**, etc.: These units of volume can be remembered because a cube has volume—so that a "cubic" anything is the volume measured using a certain unit to multiply the length, width, and height.

**Dark energy**: The universe contains a mysterious anti-gravity pressure that is responsible for the acceleration of the expanding universe in which we live. It may be associated with the vacuum of space and would, thus, grow with time as the universe grows. Its origin and nature remain unknown, but its effect can be measured and duly represented in Einstein's equations of general relativity.

**Dark matter**: A hypothetical form of matter that accounts for 90 percent of all the gravity in the universe. We have never seen dark matter but infer its presence from its gravitational effects on ordinary matter.

**Density**: A measure of how tightly packed is the material that comprises a substance. Density is formally the mass of an object divided by its volume.

**Doppler shift**: Named for the 19th-century German physicist Christian Doppler, who first measured the change in the pitch (frequency) of a sound as the sound-emitting object approaches or recedes from the listener. This shift in frequency was determined to be a general phenomenon for any form of wave.

**Drake equation**: A means to divide the overall probability of finding life in the galaxy into a set of simpler probabilities that correspond to our data and our preconceived notions of the cosmic conditions suitable for life as we know it. In the end, you are left with an estimate for the total number of technologically proficient civilizations in the galaxy with which you might communicate with radio waves or by some other technology-based methods.

**Dust cloud**: Gas clouds in interstellar space that are cool enough for the slow-moving atoms to combine and form large, complex molecules.

**Ecosystems**: A community of organisms and the environment in which they live make up an ecosystem.

**Electron**: The common subatomic particle that is negatively charged. Found in equal numbers with the positively charged protons throughout the universe.

**Elements**: The basic constituents of all matter. All matter in the universe is composed of 92 elements that range from the smallest atom, hydrogen (with one proton in its nucleus), to the largest naturally occurring element, uranium (with 92 protons in its nucleus). Although trace amounts of larger elements have been found in mines, elements larger than uranium are produced in laboratories.

**Endothermic**: In nuclear physics or in chemistry, if less energy is released in a reaction than was available at the start, the reaction is called endothermic. (See **exothermic**.)

**Escape velocity**: A special speed on all planets, stars, or anything with gravity at which a tossed object will never return. This speed is defined as the escape velocity. For all speeds less than escape velocity, the tossed object will return.

**Event horizon**: The poetic name given to the bounding region around a black hole within which light cannot escape. It may be defined as the "edge" of a black hole. This term is also applied to the visible edge of the universe.

**Exobiology**: The study of life elsewhere in the universe.

**Exothermic**: In nuclear physics or in chemistry, if more energy is released in a reaction than was available at the start, the reaction is called exothermic. (See **endothermic**.)

**Fission**: The splitting of larger atoms into two or more smaller atoms. If this occurs with atoms larger than iron, then energy is released. This is the source of energy in all present-day nuclear power plants. Also called *atomic fission*.

**Fusion**: The combining of smaller atoms to form larger atoms. If this occurs with atoms smaller than iron, then energy is released. The primary energy source for the world's nuclear war arsenals and for all stars in the universe is fusion. Also called *thermonuclear fusion*.

**Galaxy**: A system of typically billions of stars, gas, and dust that share a common center of gravity. Galaxies are the primary organization of visible matter in the universe.

**Gas cloud**: Clouds of hydrogen, helium, and trace amounts of heavier elements; the primary components of interstellar space in the disk of spiral galaxies.

**General relativity**: Introduced in 1915 by Albert Einstein, it forms the natural extension of *special relativity* into the domain of accelerating objects. It is a modern theory of gravity that successfully explains many experimental results that were not otherwise explainable in terms of Newton's theory of gravity from the $17^{th}$ century. Its basic premise is the *equivalence principle*, whereby a person in a spaceship, for example, cannot distinguish whether the spaceship is accelerating through space or whether the spaceship is stationary in a gravitational field that would produce the same acceleration. From this simple, yet profound, principle emerges a completely reworked understanding of the nature of gravity. According to Einstein, gravity is not a force in the traditional meaning of the word. Gravity is the curvature of space in the vicinity of a mass. The motion of a nearby object is completely determined by its velocity and the amount of curvature that is present. As counterintuitive as this sounds, it explains all known behavior of gravitational systems ever studied, and it predicts a myriad of even more counterintuitive phenomena that are continually verified by controlled experiment. For example, Einstein predicted that a strong gravity field should warp space and noticeably bend light in its vicinity. It was later shown that starlight passing near the edge of the Sun (as seen during a total solar eclipse) is displaced from its expected position by an amount in exact accord with Einstein's predictions. Perhaps the grandest application of the general theory of relativity involves the description of our expanding universe, in which all of space is curved from the collected gravity of hundreds of billions of galaxies. An important and currently unverified prediction is the existence of *gravitons*, or *gravity waves*. These are the particles of gravity that communicate abrupt changes in a gravitational field, such as is expected in a supernova explosion.

**Heliocentric**: Sun-centered. (Compare *geocentric*: Earth-centered.)

**Impact energy**: The energy released when an asteroid or comet strikes the surface of a planet or any other cosmic object. In the collision, the impactor's kinetic energy is passed entirely to the object.

**Impact rate**: The rate at which a planet has experienced impacts from celestial objects of a predetermined size.

**Impact record**: The accounting of all confirmed impacts on a planet's surface.

**Isotope**: A chemical element with fewer or more neutrons in its nucleus than is common in nature. All isotopes have the same chemical properties but different nuclear properties.

**Kelvin temperature scale**: Named for Lord Kelvin of the mid-19th century. He invented the scale where the coldest possible temperature is, by definition, 0 degrees. Its increments are the same as the Celsius scale. On the Kelvin scale, water freezes at 273.16 degrees and boils at 373.16 degrees.

**Killer asteroids**: Asteroids large enough to cause the mass extinction of many or most of the species of a planet. A killer asteroid on Earth would, at a minimum, destroy all civilization.

**Kinetic energy**: The energy of an object from being in motion. Mass also contributes to kinetic energy. For example, if a more massive object (such as a truck) moves with the same speed as a less massive object (such as a tricycle), then the more massive object will have more kinetic energy.

**Latent heat**: The heat that is either absorbed or released by a substance whose physical state has changed, for example, when water changes from a liquid to a solid.

**Law**: When a theory about a reoccurring natural phenomenon has been tested many times and not disproved, that theory becomes a law, like Newton's laws or Kepler's laws.

**Light year**: The distance light travels in one year. At the speed of light, this distance equals 5,800,000,000 miles.

**Long-period comets**: Comets on elongated orbits around the Sun, with periods of 200 years or more.

**Luminosity**: A measure of the rate of energy output for an object, such as a star.

**Magnetic field**: Moving charged particles are the sole producers of magnetic fields. These fields are regions of space that supply a force to other charged particles in the area. All magnets have two poles, which are often called *north* and *south*. If you represent a magnetic field with imaginary lines, then all lines form complete loops that extend through both magnetic poles.

**Mass**: A measure of an object's material content. For example, a locomotive that is weightless in space has no less mass than a locomotive that weighs 100 tons on Earth. Note also that mass makes no reference to size. A beach ball is large, but it is certainly not massive. An anvil is massive, but it is certainly not large.

**Microlensing**: Einstein's general theory of relativity describes the curvature of space-time in the vicinity of a mass. For high-mass objects, light from a distant object can be split into several paths, giving multiple images of the same object. This effect, known as gravitational lensing, also works with smaller, planet-sized masses, but in these cases, which are called microlenses, the multiple background images do not split completely. Instead, the images fall on top of one another, brightening the light of the distant object.

**Molecular cloud**: Gas clouds that are cool enough for molecules to form. Because these clouds tend to be very dense, they are the most likely places for the onset of star formation.

**Molecule**: A chemical combination of elements that normally has very different properties from its constituent parts. For example, sodium and chlorine will kill you in a high enough dose. Together, as the molecule sodium chloride, they become ordinary table salt.

**Molten**: In geology, this term is often used to describe melted rock. More generally, however, it is used to describe any thick liquid that is normally a solid.

**Momentum: angular, linear**: The tendency of an object to remain rotating (angular) or to remain in motion in a straight line (linear). Momentum is one of the "conserved" quantities in nature; in a closed system, momentum remains unchanged.

**Neutron**: A particle in the nucleus of all atoms (except for normal hydrogen). It is slightly more massive than the proton and contains no electric charge.

**Neutron star**: The tiny remains (less than 20 miles in diameter) of the core of a supernova explosion. It is composed entirely of neutrons and is so dense that it is equivalent to cramming 2000 ocean liners into a cubic inch of space.

**Northern lights**: See **aurora**.

**Nucleus**: The central region of an atom that contains protons and neutrons.

**Oblate spheroid**: A sphere that is squashed into a shape not unlike a hamburger. (Compare *prolate*.)

**Panspermia**: The idea that life on Earth may have been delivered or seeded by asteroids or comets carrying microbes on or in them when they collided with the young Earth.

**Parallax**: The change in position of a star or any other object that appears to occur just because your point of view has shifted. For example, your thumb held at arm's length will appear to be aligned with a different part of the background when you look with your left eye, then with your right eye.

**Period**: Usually, the time for an object in orbit to complete one orbit. The period of the Earth is one year.

**Periodic table of elements**: A sequence of every known element in the universe arranged by increasing number of protons in their nuclei.

**Photon**: Massless particle of light energy. Its energy determines the part of the spectrum where it would be detected. High-energy photons are gamma rays; medium-energy photons are visible light; low-energy photons are radio waves.

**Plane**: A conceptual region of space that is broad and flat. It is commonly used as a reference to orbits and orientations. For example, "The Earth's axis is tilted 23½ degrees from the plane of the solar system."

**Planetary nebula**: One of the few misnomers in astronomy, a planetary nebula is the gaseous remains of a dying red giant star. This photogenic phase of a star's life is short-lived, but nearly every star passes through this evolutionary stage, making them a common sight in the galaxy.

**Plasma**: An extremely hot gas (like a normal star) in which most of the gas atoms' outer electrons have been stripped away, leaving a charge-filled cloud that responds to magnetic fields. It is sometimes called the *fourth state* of matter.

**Potential energy**: The energy content that an object has by virtue of its chemical configuration or its position in space. For example, trinitrotoluene (TNT) has enormous chemical potential energy, and water in a dam or at the top of a waterfall has enormous gravitational potential energy.

**Primordial**: Generally refers to the chemical or physical conditions that existed at the formation of the Earth, Sun, galaxy, universe, and so on.

**Prolate spheroid**: A sphere that is squished into a shape not unlike a hot dog or an American football. (Compare *oblate*.)

**Proton**: The positively charged particles found in the nucleus of every atom. The number of protons in a nucleus defines the atom. For example, the element that has one proton is hydrogen. The element that has two protons is helium. The element that has 92 protons is uranium.

**Quantum foam**: The field of quantum mechanics describes the behavior of matter on its smallest scales. General relativity describes the curvature of space and time in the vicinity of matter. During the early universe, the entire cosmos was the size of an atom, forcing a marriage between quantum mechanics and general relativity that yielded a churned, tangled structure to space-time known as *quantum foam*.

**Quasar**: The small, but extremely luminous core of a galaxy where a super-massive black hole dines upon stars and gas clouds that drift too close. The swirling matter forms a disk that funnels down to the hole, radiating copiously along the way.

**Radiation**: Any form of light—visible, infrared, radio, and so on. In this nuclear age, however, it has come to mean any particle or form of light that is bad for your health.

**Rarefied**: Thin and wispy. Used almost exclusively to describe vacuous gases.

**Red shift**: Lengthening of the measured wavelength of light due to the motion of the light-emitting object away from you. Because motion is relative, this shift also occurs if you are in motion away from the object.

**Redshift** (one word): The general term used to describe the red-shifted spectra of nearly all galaxies in the universe. This universal redshift is primary evidence for our expanding universe.

**Relativity**: General term used to describe Einstein's special and general theories of relativity.

**Revolution**: Motion around another object. For example, the Earth revolves around the Sun. Often confused with *rotation*.

**Rotation**: The spinning of an object on its own axis. For example, the Earth rotates once every 23 hours and 56 minutes.

**Singularity**: When a very high-mass star collapses, no known force will stop it. Its material becomes smaller and denser and eventually becomes a black hole. Within the black hole, however, the material continues to collapse without limit, leading to a single point of zero size and infinite density, the singularity. Assuming that such a thing is impossible, this is evidence for the incompleteness of Einstein's general theory of relativity.

**Solar maxima/minima**: The Sun undergoes an 11-year cycle of activity, as measured by the count of sunspots that move across its surface. When sunspots are peaking, the Sun is at solar maximum. When sunspots go away, the Sun is at solar minimum.

**Space Interferometry Mission (SIM)**: A proposed space mission designed to make extremely accurate measurements of the positions and distances of stars in the galaxy, leading to the first direct measurements of extrasolar planets around other stars.

**Space-time**: The mathematical combination of space and time that treats time as a coordinate with all the rights and privileges accorded space. The special theory of relativity demonstrates that nature is most accurately described using a space-time formalism. It simply requires that all events are specified with space *and* time coordinates.

**Special theory of relativity**: First proposed in 1905 by Albert Einstein, it offers a renewed understanding of space, time, and motion. The theory is based on two *principles of relativity*: (1) The speed of light is constant for everyone no matter how you choose to measure it, and (2) the laws of physics are the same in every frame of reference that is either stationary or moving with constant velocity. The theory was later extended to include accelerating frames of reference in the general theory of relativity. The two principles of relativity assumed by Einstein have been shown to be valid in every experiment ever performed. Einstein extended the relativity principles to their logical conclusions and predicted an array of unusual concepts that include the following:

- There is no such thing as absolute simultaneous events. What is simultaneous for one observer may have been separated in time for another observer.

- The faster you travel, the slower your time progresses relative to someone observing you.

- The faster you travel, the more massive you become, so that the engines of your spaceship are less and less effective in increasing your speed.

- The faster you travel, the shorter your spaceship becomes—everything gets shorter in the direction of motion.

- At the speed of light, time stops, you have zero length, and your mass is infinite. Upon realizing the absurdity of this limiting case, Einstein concluded that you cannot reach the speed of light.

Experiments invented to test Einstein's theories have verified all of the above predictions precisely. An excellent example is provided by particles that have decay *half-lives*. After a predictable time, half are expected to decay into another particle. When these particles are sent to speeds near the speed of light (in particle accelerators), the half-life increases in the exact amount predicted by Einstein. They also get harder to accelerate, which implies that their effective mass has increased.

**Spectral type**: Any one of several lettered designations that indicate the temperature of a star. In order from hottest to coolest the designations are: O, B, A, F, G, K, M. Historically, stars were classified solely according to features in their spectra. Letters were assigned, in order through the alphabet, to classes of stars. Later, this method proved to be less useful than a classification scheme based on temperature. Many stellar classes were dropped, and some were joined with others. What remains is a hodgepodge letter sequence that is the darling of mnemonic writers.

**Spectrum**: Light after it has separated into its component wavelengths. The human eye detects wavelengths by colors.

**Sphere**: The only 3-D shape in which every point on its surface is the same distance from its center.

**Strong force of nature**: One of four principal forces of nature. One which binds together neutrons and protons to make atomic nuclei.

**Sunspots**: Small circular regions of the Sun's surface that are somewhat cooler than the surrounding areas. This temperature contrast makes sunspots appear dark against their brighter background. They move with the Sun's surface and tend to avoid the polar and equatorial regions. Sunspots commonly travel in pairs because of their association with magnetic fields. They come and go in "cycles" that define the 11-year period of solar activity. The average sunspot is about two or three times larger than the Earth.

**Supernovae**: The explosion of a high-mass star that has just run out of fuel. After having fused hydrogen to helium, helium to carbon, and so forth until iron, which is endothermic in all reactions, the star collapses under its own weight, then explodes with a luminosity that rivals the brightness of the entire galaxy of stars that surround it.

**Surface tension**: The tension at the surface of a liquid caused by the attraction of the molecules within the liquid toward one another. Surface tension is apparent in an overfilled glass of water, where the liquid surface actually rises past the rim of the glass.

**Telescope** (gamma, x-ray, ultraviolet, optical [visible], infrared, microwave, radio): Astronomers have designed special telescopes and detectors for each part of the spectrum. Most parts of this spectrum do not reach Earth's surface. To see the gamma rays, x-rays, ultraviolet light, and infrared light emitted by many cosmic objects, these telescopes must be lifted into orbit above the absorbing layers of the Earth's atmosphere. Although all the telescopes are of different design, they do share three basic principles: (1) They collect photons, (2) they focus photons, and (3) they record the photons with some sort of detector.

**Terrestrial Planet Finder (TPF)**: A proposed NASA mission designed to locate and image Earth-like planets around other stars. The brighter planets will be targeted for follow-up spectra where the signature of surface life may be found in the atmospheres of the planets themselves.

**Theory**: A general principle that is widely accepted and is in accordance with observable facts and experimental data.

**Thermodynamics**: The study of heat as it interacts with other forms of energy and matter.

**Thermal energy**: The energy contained in an object (solid, liquid, or gaseous) by virtue of its atomic or molecular vibrations. The average kinetic energy of these vibrating atoms and molecules defines temperatures.

**Thermonuclear**: Any process that relates to the behavior of the atomic nucleus in the presence of high temperatures.

**Tidal forces**: The difference in gravity from one side of an object to another, creating a sustained stress that elongates the object in the direction of the source of gravity. In extreme cases, the tidal force will exceed the binding forces of the object, forcing it to break apart.

**Tunneling**: An effect in quantum mechanics that has no analog in classical physics. Even though a particle might not have enough energy to cross an energy barrier, tunneling is the assured probability that the particle will get there in any case, similar to the example of a car driving through a mountain instead of over it. The protons engaged in thermonuclear fusion in the Sun's core require tunneling to undergo fusion.

**Wavelength**: The length of a repeating component of a wave. A very useful term that applies to sound, light, trucks in a convoy, ripples on the surface of water, and so on.

**White dwarf**: The end-state in the life of an intermediate-mass star where its nuclear fuel supplies have run out. In the process, the outer layers of the star expand to make a red giant, while the inner layers collapse to form a hot, dense, Earth-sized ball of matter—the white dwarf—laid bare as the outer layers float away into space.

# Biographical Notes

**Copernicus, Nicolaus** (1473–1543), Polish astronomer: The greatest astronomer since Hipparchus, he is generally credited with restoring the idea of a Sun-centered, heliocentric universe. For nearly 1500 years, the Earth-centered, geocentric universe had been predominant.

**Doppler, Christian** (1805–1853), Austrian physicist: A distinguished physicist in his day, Doppler is best known for the spectral effect that bears his name. When an object in motion emits a wave of anything (such as sound or light), the natural frequency of the wave is increased if the object is approaching you and decreased if the object is receding. This general principle, the *Doppler shift*, reveals itself in train whistles, racecars, and the expanding universe.

**Eddington, Sir Arthur** (1882–1944), English astrophysicist: Eddington married his tandem knowledge of physics and astronomy to become the first astrophysicist. A brilliant scientist with a tireless interest in the latest and most important problems of the day, Sir Arthur was well known for conducting the first confirming measurements of the curvature of space-time, as predicted by Albert Einstein in his modern theory of gravity, which is better known as the general theory of relativity. Sir Arthur also attempted to deduce the nature of stars and other cosmic phenomena from physical principles. Although not always correct, his efforts reliably stimulated further research by others.

**Einstein, Albert** (1879–1955), German-American physicist: Einstein's contributions to our physical understanding of the universe rival only those of Isaac Newton. In 1905, he proposed his revolutionary special theory of relativity, where space and time are conjoined. This conceptual framework allowed him to make counterintuitive predictions about the mass, the flow of time, and the physical dimensions observed of an object as its speed approaches the speed of light. Followed by his 1916 theory of gravity—the general theory of relativity—Einstein interpreted gravity as the curvature of space-time through which matter falls, rather than as a conventional force that acts at a distance. To date, all reliable experiments that have ever been conducted have confirmed the predictions of relativity.

**Galilei, Galileo** (1564–1642), Italian physicist: Although not the inventor of the telescope, Galileo may have been the first to look up with it. What lay before him was a garden of cosmic knowledge that permanently altered the landscape of scientific thought. His discoveries ranged from simple observations that the Moon's surface is not smooth (as presupposed) to the fact that Earth cannot be the center of all motion because Jupiter has a set of moons all to itself. His heretical findings and his relentless ego got him in trouble with the Catholic Church. He was found guilty and, to avoid torture, was forced to sign a confession that renounced his data. Galileo (rather, his corpse) was found innocent of all charges somewhat later (350 years) by Pope John Paul II.

**Gamow, George** (1904–1968), Soviet-American physicist: Gamow was a distinguished physicist at a time when human understanding of the atom and the universe took fundamental leaps. In 1946, he proposed what came to be known as the *Big Bang* model of the universe, which came with testable predictions of the abundance of heavy elements and of a background remnant of microwaves from the original explosion.

**Hertz, Heinrich Rudolf** (1857–1894), German physicist: Hertz demonstrated that radio waves are just another form of electromagnetic waves, akin to visible light, thus enabling the intellectual unification of previously disjointed forms of energy. The familiar unit of electromagnetic frequency is named in his honor.

**Hubble, Edwin Powell** (1889–1953), American astronomer: Among many seminal contributions to observational astronomy, Hubble discovered the expanding universe in 1929. Law and boxing were two early interests of his before he turned to the heavens.

**Humanson, Milton Lasell** (1891–1972), American astronomer: Best known for his work on the 100-inch telescope at the Mt. Wilson Observatory and, later, the 200-inch telescope at Mt. Palomar, Humanson obtained spectra of galaxies in the late 1920s and for decades to follow. His data enabled Edwin Hubble to further extend his discovery that distant galaxies recede faster than nearby ones.

**Kant, Immanuel** (1724–1804), German philosopher: Among astrophysicists, Kant is best remembered for having proposed in a 1755 essay the "nebular hypothesis" to explain the origin of the solar system. Kant suggested that a large spinning gas cloud would flatten as it collapsed under its own gravity. A large central nucleation would form the Sun, while smaller nucleations would form the planets. Although various modifications to this suggestion have been required over the years, the basic idea and scenario are correct. Extending this idea to the entire galaxy, Kant also supposed that the fuzzy "stars" in the sky were other galaxies—island universes distinct from our own, an idea later confirmed by Edwin Hubble in 1929.

**Kepler, Johannes** (1571–1630), German mathematician and astronomer: Kepler proposed the first truly predictive mathematical theory of the universe through his laws of planetary motion. Isaac Newton would later show that Kepler's laws are derived easily from more basic theories of gravity.

**Laplace, Pierre Simon** (1749–1827), French mathematician and astronomer: Duly famous in the annals of astronomy for many reasons, Laplace, most notably, updated Isaac Newton's laws of gravity to allow for the hard-to-predict multiple effects of many sources of gravity acting simultaneously. In what is today called *perturbation theory*, Laplace's technique allowed one to calculate planetary orbits with unprecedented precision. In the face of this enlightened understanding of celestial motions, Napoleon Bonaparte once commented to Laplace that there was no mention of God in his book, whereupon Laplace replied, "Sir, I have no need of that hypothesis." Laplace also postulated the existence of an object with such high gravity that light might not escape, and he proposed independently that the system of planets owes its origin to a collapsing, flattening, rotating gas cloud.

**Lippershey, Hans** (c. 1570–1619), Dutch optician: Credited with being the first person to assemble two lenses in such a way that objects appear closer to the person who looks through them. This invention is known as the telescope.

**Lord Kelvin**: See **Sir William Thomson**.

**Michelson, Albert** (1852–1931), American physicist: Best known for his development of the *interferometer*, which is an extremely sensitive optical device that can be used to measure, among other things, the speed of light, to unprecedented precision. Teamed with Edward Morley in 1887, he demonstrated that the speed of light was independent of the direction that Earth moved through the ether, thus casting serious doubt on its existence as a medium through which

©2003 The Teaching Company Limited Partnership

light must travel. In 1907, Michelson was the first American to ever receive the Nobel Prize.

**Morley, Edward Williams** (1838–1923), American chemist: See **Albert Michelson**.

**Newton, Isaac** (1642–1727), English physicist: A famous quote from Newton proclaims that if he can see farther than other men it is because he stands upon the shoulders of giants who came before him. This may have indeed been true (especially if the giants are Copernicus, Kepler, and Galileo), but the real secret to his distant vision might simply have been that he was surrounded by intellectual midgets. In spite of this, Newton spent his most scientifically productive years alone, during which he discovered the laws of gravity and many laws of optics and invented calculus. Surely one of the greatest intellects ever to walk the Earth, a statue of him in Cambridge, England, proclaims: "Of the human species, the brilliance of Isaac Newton reigns supreme."

**Slipher, Vesto Melvin** (1875–1969), American astronomer: An accomplished astronomer who is best known for obtaining spectra of spiral nebulae that enabled Hubble to conclude that the spiral nebulae were indeed entire galaxies external to our own Milky Way and that nearly all were moving away from us.

**Thomson, William** (Baron Kelvin of Largs) (1824–1907), British physicist: A precocious lad, William Thomson graduated from the University of Glasgow at age 10. He became a major contributor to human understanding of the electromagnetic force and the study of thermal energy, better known as *thermodynamics*. The Kelvin absolute temperature scale, where zero degrees is the coldest possible temperature, is named in his honor.

# Bibliography

Adams, Fred, and Greg Laughlin. *Five Ages of the Universe*. New York: Free Press, 1999. An intriguing account of the birth, life, and death of the universe from its earliest moments to its most distant future. Scenarios include many that are catastrophic or decidedly unpleasant for Earth dwellers.

Altshuler, Daniel R. *Children of the Stars*. Cambridge: Cambridge University Press, 2002. An updated, popular account of the formation of heavy elements in the core of high-mass stars and their subsequent return to the interstellar medium via supernova explosions.

Anonymous. *The Bible According to Einstein*. New York: Jupiter Scientific Publishing Co., 1997. A remarkable book, conveying all of modern science in a form and narrative inspired by the Judeo-Christian Bible. Humorous in places but always informative. Again, attempting to synthesize the relevant science that informs our modern understanding of the universe and the events that shaped it.

Dorminey, Bruce. *Distant Wanderers: The Search for Planets Beyond the Solar System*. New York: Springer-Verlag, 2001. An account of the who, what, where, why, and how of the discovery of planets beyond the solar system.

Feynman, Richard P. *The Character of Physical Law*. Cambridge: MIT Press, 1973. A remarkably succinct yet readable account of the way nature works, including important observations on the universality of physical law.

Goldsmith, Donald, and Tobias Owen. *Search for Life in the Universe*. Mill Valley, CA: University Science Books, 2001. More a textbook than a general reader, this book (3rd ed.) remains one of the definitive surveys of the subject.

Guth, Alan H. *The Inflationary Universe*. Reading, MA: Addison-Wesley, 1998. Written by one of the pioneers of the inflationary Big Bang, this account remains the best introduction to this peculiar episode in the early universe, where space-time expanded faster than the speed of light.

Lewis, John S. *Mining the Sky*. Reading, MA: Addison-Wesley, 1996. An alternative perspective that treats asteroids as natural resources to be exploited. In doing so, we may just learn to deflect them out of harm's way.

————. *Rain of Iron and Ice*. Reading, MA: Addison-Wesley, 1996. If you are not yet scared of getting slammed by an asteroid, perhaps you should have a look at this account of our future on Earth.

Margulis, Lynn, and Dorian Saga. *What Is Life?* New York: Simon & Schuster, 1995. You can't look for life unless you understand what you are looking for.

McKay, D. S., et al. *Search for Past Life on Mars*. Washington, D.C.: American Association for the Advancement of Science, 1996. The closest the subject of exobiology has ever come to real data has been our ongoing attempt to probe the Martian soils for life, including the extensive analysis of a meteorite from Mars that might harbor evidence for a long-extinct biota on the Martian surface.

Schulman, Eric. *A Briefer History of Time*. New York: W.H. Freeman & Co., 1999. In the spirit of Lecture Nine, this book attempts to encapsulate the most important events in the cosmic history into a single comprehensive narrative.

Shipman, Harry L. *Black Holes, Quasars, and the Universe*. Boston: Houghton Mifflin Co., 1976. There's no better title or treatment for a book on the best of the bizarre in the cosmos.

Smoot, George, and Davidson, Keay. *Wrinkles in Time*. New York: William Morrow & Co., 1993. Reporting on earlier data that are now surpassed by the results of the WMAP satellite, this book contains a good introduction to Big Bang cosmology, blended with some of the trials and tribulations of how big science gets it done.

Spitzer, Lyman. *Searching Between the Stars*. New Haven: Yale University, 1982. Sptizer was the world's expert in all that happens between the stars. This book, slightly more advanced than the average treatment intended for the public, contains a broad account of why we should care about the life cycle of stars and how they enrich the clouds out of which subsequent generations of stars form.

Thorne, Kip. *Black Holes and Time Warps*. New York: W.W. Norton & Co., 1994. One of the world's experts tackles this challenging subject with ample anecdotes from a scientific, as well as a historical, point of view.

Tyson, Neil deGrasse, Charles Liu, and Robert Irion. *One Universe: At Home in the Cosmos*. Washington, D.C.: Joseph Henry Press, 2000. Because of the book's themes and organization, rarefied objects and related phenomena take on an entire category in the section on matter and its extremes. (Visit http://www.nap.edu/oneuniverse.)

Vershuur, Gerrit, L. *Impact: The Threat of Comets and Asteroids*. Oxford: Oxford University Press, 1997. If you are not yet scared of getting slammed by an asteroid, you should have a look at this account of our future on Earth.

Voit, Mark. *Hubble Space Telescope: New Views of the Universe*. New York: Harry N. Abrams, 2000. One of many books that attempt to bring the splendor of the Hubble's universe down to Earth.

Ward, Peter D., and Donald Brownlee. *Rare Earth*. New York: Springer-Verlag, 2000. Argues strongly for a fertile Earth that sits, possibly uniquely among planets, as a haven for complex life, including humans. The conditions that give rise to this circumstance offer insight to the beginning and end of the world.

Wheeler, J. Craig. *Cosmic Catastrophes*. New York: Cambridge University Press, 2000. A readable book from a well-known astrophysicist who happens to be a long-time enthusiast of the most energetic phenomena in the cosmos.

**Internet Resources:**

http://heritage.stsci.edu/. Hubble Telescope online archives, including a site showing the most beautiful full-color images ever obtained of the universe, all capturing some remarkable astrophysical object, phenomenon, or idea.